S0-BXH-743

THEY BLEW OUR WEATHER

THEY BLEW OUR WEATHER

FREDERICK CLARE

Exposition Press   Smithtown, New York

A percentage of the royalties from the sale of this book will be donated to the American Kidney Fund.

First Edition

ISBN 0-682-49824-6

Printed in the United States of America

To those in and out of government who willingly contributed information, photographs, graphs, drawings, and research time—so much data—to make this semi-documentary available to people everywhere

and to

Peter Bergel

Jan. 20, 1989

[illegible]

Contents

Acknowledgments

I gratefully acknowledge the publishers, authors, photographers, and artists who have kindly allowed me to use their material. To the many friends both in and out of government who have assisted me in the very difficult research over the years, many thanks. I am especially grateful to:

The U.S. Department of Defense; the U.S. Department of Energy, Jack S. Schneider, still-photo officer of Public Affairs; James S. Cannon, OPA; Philip A. Garon, OPA; D. Tank and Joan Walter of NASA, 400 Maryland Avenue S.E., Washington, D.C.; Martha Casey, News Services Division, USEPA; Ron Surface of the U.S. National Weather Service; the U.S. Naval Photographic Center; the National Environmental and Satellite Service; Timothy Nolan, Jr., of the Department of Health, Education and Welfare, Public Health Service, Center of Disease Control; OEPBS channels for the "Nevada Fallout" presentation. Ann B. Strong and J. Michael Smith, of the Eastern Environmental Radiation facility, and Raymond H. Johnson, Jr., of the Environmental Analysis Division (AW-461), for their excellent report about nuclear bomb testing by the People's Republic of China, published by the EPA in 1976.

Samuel Glasstone and Philip J. Dolan, editors of the government study *The Effects of Nuclear Weapons,* third edition, 1977, a remarkable compilation of data and photos which furnished many illustrations, graphs, and technical data for this book.

George L. Toombs, supervisor, Environmental Radiation Surveillance Program, Radiation Control Section for the State of Oregon, and his crew. David Hopkins, researcher, Oregon Center for Health Statistics, and his crew, for the excellent graphs and prompt delivery of the information included herein; John L. Isbister, M.D., Disease Control Officer, Department of Public Health, State of Michigan, for information on the problems of the Hemlock area of Michigan; the University of Minnesota, the Underground Space Center 11, Mines and Metallurgy, Minneapolis, Minnesota 55455, for the photos and drawings; the Department of Public Health, State of Michigan, Dr. Maurice S. Reizen, director.

National Archives and Record Service and the National Climatic Center, Satellite Data Services Division, both of Washington, D.C.; Readers Digest Association, Inc., United Press International; *Los Angeles Times;* Chilton Publications; Flick-Reedy Corporation, George Buctel, graphic design; Ellis Lucia, freelance writer and photographer, author of *The Big Blow,* Portland, P.O. Box 11507, Oregon 97211; Oregon Historical Society, 1230 S.W. Park Avenue, Portland, Oregon; Dempsey Center, 1427 S.E. 182, Portland, Oregon 97233; *Iron Age* magazine and its editor, Gene Beaudet, Chilton Way, Radnor, Pennsylvania 19089; Gertrude E. Myers, 3109 S.E. Walta, Vista Drive, Milwaukie, Oregon, for photograph of rays in Oregon's October storm; Mr. and Mrs. Finnell for their *Teacher's Edition of the Bible,* translated from the original tongues, 1889; Evelyn Bastian, for her biblical research, Portland, Oregon; Pastor R. Scott Harkness, the Reverend Max Wills, and Rosemary Glatfelter, historian, all of the Methodist Church; the Reverend George E. Vandeman, author of *It Is Written* and *The Day Uranium Went Critical;* Lawrence Maxwell, editor, *Signs of the Times,* Seventh Day Adventists Publication; Pastor Fresk and others of his flock.

Wide World Photos, Inc., for photo Blizzard of 1888 and tanks at test site; Woodfin Camp, Inc., for photo of Ten Commandments; The *Plain Truth* magazine, 300 West Green Street, Pasadena, California; Mrs. E. M. Howard, proofreader; John E. Barnard, Jr., A.I.A., architect for "Going Under to Stay on Top" ecology house; William Juan, Photo Laboratory Supervisor, Los Angeles, California; Barbara A. Shattuck, Illustrations and Requests, *National Geographic* magazine.

Ross Hamilton of the *Oregon Journal*'s photo division, Portland, Oregon, for his fine photo of the Mount St. Helens eruption; Pat McLelland, 3634 S.E. Ogden Street, Portland, Oregon 97202, for the fine sketches which enabled me to better convey the real meaning of this book; Jean Dufour, 19 r d, Vioux Sextier, 84,000, Avignon, France, for his photo of a Holy Land dust or sand storm.

The Institute of Marine Biology of the University of Oregon for the research data and photos of the whale strandings on the southern coast of Oregon, especially Dr. D. H. Varoujean and Melomie Rice.

Introduction

They Blew Our Weather is about just that. How *they* have blown huge holes in that great ocean of air above us and how the extreme cold and sometimes extreme heat of outer space are dumped back to earth from miles away (weather bureau records show extreme heat and cold changes of as much as fifty degrees in minutes, with time, date, and place)! How and why sudden storms come straight down out of the sky, with no warning whatever! Why the superpowers suddenly ceased atmospheric testing! Here also are numerous discoveries made by moon rockets that have seldom if ever been publicized, new discoveries about the ocean of air above us; about the extremes of cold and hot the higher you go, until there is a fiery furnace! How underwater detonations have poisoned our oceans along with other radioactive wastes. It tells why whales beach themselves. It also tells who *they* are around the world, not in just the good old U.S.A.!

The history of great civilizations and the cause of their downfall should be a warning to us. It is all entangled with politics, ecology, deadly invisible rays and radiation that boggle our minds and poison our bodies. Thirty-six years after the first bomb was dropped, research reveals what is happening from exposure to rays, radiation, and fallout, how much is fatal, internally and externally.

Insects seem not to be affected. Giantism genes seem to be stimulated. Fauna, flora, mankind, things in the seas and rivers are affected by direct or delayed fallout from bombs and volcanoes! The problems of medical diagnosis of radiation poisoning, some of which are identical to virus and disease symptoms, need intensive study now. Research will find a way. The deadly effects of plutonium may already be conquered!

They Blew Our Weather shows how underground living can be a new way of life; how fallout, fire, and tornado shelters can be nice places to live. Montreal, Canada, has a complete underground city, rapid transit and all! The Russians have similar shelters but built with an atomic war in mind. So have the Chinese. The Russians aren't dumb. An atomic war is the last thing they want. They know it's world suicide!

Acts of God and Mother Nature are clearly shown here to be different, contrary to insurance companies' definitions. This book lists prophecies from the Bible that have come true, prophecies that are happening now and those that are going to happen. It shows what has been, what's going on now, and what can be, if we want it!

There is no better way to find out quickly what this book is all about than to read the true story of an American hero, the nuclear physicist Louis Slotin, who gave his life for his co-workers and his country!

1

Manmade, Natural, and Prophesied Happenings

THE DAY URANIUM WENT CRITICAL

A Day to Remember
by
George E. Vandeman

Shortly after the New Year of 1946, preparations were being made to conduct atomic tests in the lagoon of the Western Pacific atoll of Bikini, a U.S. trust, about 15 degrees, north of the equator. A young scientist, Louis Slotin, was carrying on experiments in an effort to determine how much uranium U-235 was necessary to bring about a chain reaction. This had been determined after several tests. As he was to leave for another assignment, and due to the long distance to Bikini, he felt that he should make the test once more, just to be sure.

The test consisted of bringing two spheres of uranium slowly together, until the masses became critical and then push them apart instantly with a laboratory tool to stop the reaction!

The young man pushed his luck too far that day. Just as the hemispheres became critical, the tool slipped. The room was instantly filled with a blinding, bluish haze! Everyone froze. Louis Slotin, knowing that in less than a second there would be no laboratory in existence, forced the two masses apart with his bare hands. He saved the lab and the lives of the others in the lab.

He knew he was doomed to die, but shouted to the others to remain exactly where they were when the accident happened. This was so doctors could discover the degree of radiation to which they had been exposed. He drew on the blackboard the exact positions they were in at the time of the flash. Later when he and his buddy Al Graves, who had been next

> closest to the uranium, were waiting for transportation to the hospital, young Slotin said to Al, well aware of the fate that awaited him, "Al, you'll come through all right, but I haven't the faintest chance to make it." Nine days later he died of radiation poisoning—a hero to his friends, science, and you and me. (From what we know now, it's doubtful whether the rest of them survived either.)

The Medal of Honor should have been awarded to Louis Slotin's family posthumously. So that you will know one when you see it, see Plate 1 in the color photo section. It can be found in hock shops. You can't eat it. It's not good for a night's lodging. The least Congress should do is make the recipient tax exempt for life!

OUR FREAKISH WEATHER

Attempts have been made to explain the world's freakish weather through logistical study of ocean currents; others have blamed it on highs and lows of air pressure, in connection with ocean water temperatures. Wandering jet streams have also received their share of condemnation. Why are they wandering and so difficult to locate? They used to be so dependable. Oceanographers and climatologists have tried to put the finger on why it's happening. Those who know the reason have carefully concealed it from the public. At last, *the answer is here!* What you see on the jacket photo has been going on since July 16, 1945, after Hiroshima and Nagasaki, after that, *they* began subjecting our own people to almost the same exposures! Right here in our own country, it was done! Should God forgive them, for they knew not what *they* did? Don't you believe it.

Climatologists say that an atomic bomb will have no more effect on the weather than a severe thunderstorm! Yet no atomic or fusion bomb detonated above ground can be controlled. Those in charge don't do too well controlling underground blasts either. Above ground there is the blast effect—air pollution and a fallout of silent, unseen as well as the visible radioactive particulates, deadly to all life as we know it. It's cumulative and can't be heard falling on a tin roof. However, hundreds of Nevadans have seen it, breathed it, eaten it, lived in it, and died from it! Preposterous?

Below is a transcript from the sound track of a government film entitled "Nevada Fallout."

(This film is a must to see on TV, if and when it appears on public broadcasting in your area.)

Nevada Fallout

VICTIM NO. 1: The glow and brilliance of the landscape was so ethereal; it made you feel funny. After the first one I didn't know just what was the matter with me. I lost all my hair and it never grew back. After that first time, you just accepted it because there were so many that followed. My children even had burns. At school they had a rash on them like sunburn.

NARRATOR: One hundred and twelve thousand Nevadans have been exposed to heavy radioactive fallout since 1951! Vast tracts of Nevada real estate have been scarred and dusted with nuclear debris! Five hundred times the stillness and quiet of the southern desert has been shattered in the name of national security! One hundred times the tongue of death has lashed across the desert through the boundaries of the Nevada test-site perimeter and into the valleys, mountains, ranches, and towns of the rural countryside! I'm standing about two miles from Queen City summit, on highway 25 which runs east and west across the state of Nevada. Warmsprings and Railroad Valley lie about fifty miles off to the west. In this spot here, this fence marks the northern boundary of the vast Nevada test site. The test site covers about 4,000 square miles, larger in fact than the state of Rhode Island. During the days of the atmospheric program, when the winds were blowing right, radioactive debris would be carried to the north, to this relatively uninhabited area. It was in this region here that the largest radioactive readings ever were recorded! Atomic devices were and still are detonated to the south of this northeastern flank of the Nevada test site and gunnery range. The official accumulation of radiation, shown growing here over a twenty-eight-year period, had left its footprints on these areas of the state. The clouds of radiation would *completely* engulf you. The mountains around here would block them in. You wouldn't be able to see from here to the barn when the clouds moved in!

VICTIM NO. 1: We rode through this stuff several times. It burned like a real hot sunburn; it would burn your face.

INTERVIEWER: Did you yourself get burned?

VICTIM NO. 1: Yes, everyone here at that time was burned. My father and my aunt, they got particles in their eyes which gave them quite a bit of trouble and they had to go to the doctor for it. We noticed that almost all of the birds died. At that time you couldn't find a bird in the country. One thing that stood out was that we had large numbers of canceroid cattle and burned cattle. The dogs, we had four of them, ended up with huge tumoroid growths, and all of them died.

INTERVIEWER: You took quite a number of pictures back at that time?

VICTIM NO. 1: Yes, I did, but had no luck, until I found how to shield them from the radiation. I'd take pictures and when I got 'em back they were completely exposed. It ended after talking to people that knew something about radiation. I had a bunch of lead blocks and took pictures before the cloud would actually engulf us. Then I put 'em in the icebox and shielded them with these blocks of lead around them. I got pictures then. Joe Cellini had concealed these pictures of late 1950 from the government.

NARRATOR: Here one cloud, seen belching across the desert from the south, has moved into the heart of Railroad Valley and toward the Sharp ranch at Niella, Nevada, thirty miles away. It was here that Minnie Sharp often worked bareheaded in her garden, occasionally looking up at the clouds nestling along the mountains near her home.

INTERVIEWER: Was it often that the clouds would come over into this valley?

VICTIM NO. 1: They all came over this way. We saw them all the time. They all came over. Some drifted up the other side of the valley and some up this side, but they all came into the valley. Sometimes we had some awful dirty ones. We even had a monitor out here, watching. He'd go out and register it on his geiger counter, and they kept us in the house for all day for that bomb. For five hours we never saw the sun! There was so much dust and dirt. I got sick about the same time Butch Bardoli did, and we had to go to the doctor's in Reno. I didn't have any white corpuscles and had one infection after another and couldn't fight 'em, because I didn't have enough white corpuscles in my body to fight the infections. I lost all my hair and it never did grow back. After we watched a bomb go off we saw the flash and after that the concussion.

INTERVIEWER: You actually felt it?

VICTIM NO. 1: Oh, we felt it quite definitely. Some would go off about twice. It blew our front door open and cracked the kitchen window.

INTERVIEWER: The cloud was in a clear Nevada sky, which is usually cloudless?

VICTIM NO. 1: Yes, most of the time.

INTERVIEWER: You're sure it was a radioactive cloud?

VICTIM NO. 1: No doubt about it.

INTERVIEWER: Did it have any special color?

VICTIM NO. 1: Sometimes it was pinkish, sometimes a smoky color, and sometimes a kind of gray color.

NARRATOR: Martha Laird Bardoli also lived in Railroad Valley, thirteen miles from Minnie Sharp. Martha's son Martin (Butch) was seven years old in 1955, attended a tiny school at the Fallini Ranch. Martin and several other children at the school would wait excitedly for the mushroom clouds to appear. In those days the tiny school had been unknown to the AEC.

INTERVIEWER: Neither parents nor children were warned about these early tests?

VICTIM NO. 2: No, they were not. At no time did they ever contact us, nor did they ever come to our place until Martin passed away. Martin, my son, contracted leukemia. They never tested our soil or tested our milk or the meat. We ate the vegetables from our garden the year round, and I canned a lot. Just about all the vegetables we ate came out of our gardens.

NARRATOR: In February, 1957, Martin came home from school at noon one day complaining of cramps and fever. His mother, thinking it might be appendicitis, took him to a doctor in Tonopah. There he was treated for a virus infection but his fever persisted. Later a Reno pediatrician, Dr. Palmer, diagnosed it as leukemia. He suggested at the time that Martin's problem might be fallout related. Nine months later young Martin died!

INTERVIEWER: Were there many children at the school, when he first started at the school?

VICTIM NO. 2: There were five children at the school during the time Martin attended there. Later on there was a road crew that came in. They were building a road. Some of the children from there attended the school, for just a short period of time. One of the teachers that taught school there, she got cancer and she died! My brother got cancer and died! One of the Fleme boys got cancer and died. My boy got cancer and died. My sister went blind. One of the hired men got cancer and died. My children were burnt by radiation.

NARRATOR: This is Railroad Valley. We are located right now smack dab in the middle of it. The atomic test site is located about fifty miles to the south of us. The Martin Bardolis lived here. They had a ranch not far away. Martin [their son] traveled to the school every day about thirty miles. When Martin died, Mrs. Bardoli sold the ranch and moved away. The ranch changed hands two or three times and finally the ranch has gone back to nature. No one lives there any more. After the agony of Martin's death, Mrs. Bardoli wrote letters to senators and congressmen, protesting to government officials, the agony of her son's death. One response from U.S. Senator George Malone suggested that the scare stories regarding fallout and, by implication, Martin's death, were Communist inspired! Another stated that it was a small sacrifice for people living in that area.

NARRATOR: Moving east from Railroad Valley but still in that area of fallout shadow is the wintering range for sheep ranchers from eastern Nevada and Utah. Twenty-six years ago, in the spring of 1953, some particularly dirty bombs, code names Nancy and Harry, let loose their fury in these valleys. Local people still call these the ones that got away.

VICTIM NO. 3: The first symptoms we noticed were burns on their [sheep] backs or on their heads, where there were no amounts of wool.

NARRATOR: Fallout from the tests lodged in the fleece of the sheep and settled on the range grass. Surprisingly, large concentrations of iodine-131 were found concentrated in the thyroids of the sheep—in some cases hundreds of times higher than that permitted for humans. Although well nourished and fat, the lambs were stillborn and sheep ranchers like Ken Clark suffered large financial losses—a pretty sickening feeling just being wiped out! He died from the worry of seeing everything he had worked and scraped together all his life going down the drain in such a short period of time.

It's summer now in the middle of the Nevada sheep country. The sheep have since all moved off, over in that direction there, over to the Utah border. There is plenty of evidence the sheep were here though on the ground around us. The air today is moving—hard, dry, and strong from the south, much as it might have been back in the spring of 1953, the spring of dirty Nancy and Harry.

Bill Schofield has lived near Alamo just south of the sheep range for sixty-six years. He vividly remembers the effects the dirty ones, Nancy and Harry, had on local cattle ranchers.

VICTIM NO. 4: At that particular time, the men who ran range cattle out to the west of us here, in what was called the Gold Basin area, close to the test site, brought in cattle and horses with scabs and burns and discolored hair on their backs. The evidence then was radioactive contamination.

NARRATOR: In the spring of 1953, six cattle walked away from a badly contaminated water hole near Alamo and collapsed dead! Since the cattle showed no signs of external damage, the rancher's claims against the government were ignored, although the same rancher had been compensated for sixteen badly burned horses, presumably because the damage was visible. Bob Edmonson is chairman of the Nevada Governor's Committee on Radiation Effects. He says, "There is no question that the sheep deaths were radiation caused. All of that is going to be relevant in our studies of what has happened to man."

The Department of Agriculture has been conducting tests on a long-term basis since about 1962. Studies on animals on and off the test site were made, and thyroid tests have shown iodine, 100,000 times above normal, in Reno slaughterhouses in the 1960s. Louise Whipple Acre, of Hiko, Nevada, just north of Alamo, shares a common and bitter resentment, along with the thousands of Nevada test site people, that she and her family have been used as human guinea pigs in a macabre experiment for the last *twenty-eight years!* The feeling has intensified in recent years after the death of her youngest son, Kent, of cancer.

VICTIM NO. 5: I felt that it was later that we were used as guinea pigs. . . . We accepted gratefully without complaining. I never made a complaint.

INTERVIEWER: When did your son start having problems?

VICTIM NO. 5: Well, for cancer, you mean? That was just a year before he died. That was when he was down in the St. George area, working in the dust.

NARRATOR: Her son had to work cattle in the area of heavily contaminated dust. His body was already full of what they were breathing in their valley.

VICTIM NO. 5: This seemed to aggravate what he already had.

NARRATOR: Kent was twelve years old when the dirty bomb, Harry, roared through these valleys. He was strong and healthy, a conscientious family man who received regular checkups. In his later years, he traveled repeatedly through the most heavily impacted areas of Nevada and Utah. Still his cancer was detected late. Working furiously to put his life in order, he posed for a family portrait just two months before he died. He wanted his family to remember him as a strong and happy man. While cancer and radiation are known to be related, just how, no one seems to know, or will say! The exact mechanism is still unclear.

There is a mechanism, a genetic mechanism that controls cell division. What radiation does, or any mutagen, is to override that mechanism and you get an uncontrolled division of cells. Uncontrolled division of cells is a cancer. Why does it take five years or even ten or longer for an effect to show for someone who has been exposed to radiation? Now, we know that if you look at all other carcinogens, excluding radiation, there is a lag time. Apparently, a few cells in the body are affected and the body combats them; over a time, the abnormal cells get a little better foothold and finally break out into something detectable.

Psychologists have noted a great deal of worry and fear from persons who have lived throughout the fallout area. Aaron Smith is the research coordinator for the V.A. medical center in Reno, Nevada. He says, "When an individual loses a member of the family to cancer there is upset. Depression is certainly to be expected. One of the things that I've observed that isn't anticipated is that the surviving member expects to follow shortly, from the same disorder. . . . They say, 'I know that I'll be next, because I got just as much radiation as my wife. It's just a matter of time until I die.' We have proliferations about length of life I don't think we would ordinarily find in persons who are bereaved. I think that this is one of the things that has made me suspicious that we may have as much potential for destruction in the psychological area as in the physical area. Perhaps the greatest fear among the Nevadans that I have spoken to is the fear of genetic damage or birth defects. There is some evidence that this gruesome fact has occurred in the fallout areas. On the record about questions of this sort, they are met with silence. Dominant mutations might show up immediately. Reflective mutations might remain hidden for generations before cropping out. Several ranchers have had personal problems of this sort but being personal they did not want to comment on it. Mutation can and will be passed on to future generations. There's a time bomb in Nevada soil!" [The soil of the world is also contaminated to a lesser extent, witness the following news item:

> Sofia, Bulgaria (UPI). One baby with two heads or two babies in one body was the problem facing physicians at a provincial hospital, the official Bulgarian news agency BTA said.
>
> They determined that the child, born to a twenty-three-year-old mother, is twin girls sharing one body and set of limbs, BTA said Wednesday.
>
> They have two heads, two nervous systems, two spinal columns, and two gullets. But in the common body there is only one heart, one bladder, and two kidneys, BTA said. BTA said a reporter spent fifteen minutes observing the twins.
>
> "The first impression is strange; you see a normal baby body (weighing at birth 8 pounds, 1 ounce) and two heads covered with dark hair," said reporter Sergal Nakov.
>
> "The common chest (thorax) goes up and down very frequently; breathing seems hard. In fact, these are two babies named for convenience but very adequately, Ana-Maria," he said.
>
> "The left head belongs to Ana, and the right to Maria. When the nurse tickled the left foot, Ana opened her eyes and cried. Maria continued to sleep. In a little while they tickled the right foot and Maria started making faces while Ana slept on."]*

A series of twenty-six so-called safety tests, all made before 1959, spread large quantities of highly radioactive uranium and plutonium over large areas of the state. One hot spot, centered near Eureka, Nevada, shows both uranium and plutonium in the soil at five to ten times the national average.

*Not part of the sound track.

We have information that indicated that there was a considerable amount of experimentation done with cladding, of testing weapons to see if they would go off in different configurations and didn't go critical, but sometimes they did, and in any event a variety of substances went up into the air and consequently fell across Nevada. Some of that was radioactive tungsten, plutonium, and some other exotic metals.

INTERVIEWER: Does the element plutonium pose any specific problems to people exposed to it?

NARRATOR: Prior to our entry in the nuclear age, plutonium was so rare that it was almost unknown on the surface of this planet. It is probably the most toxic compound. It has a half life of 24,000 years. Plutonium oxide has just the right-sized particles so that if you inhale it, it will lodge in the lung, so that if you've got a little bit of plutonium in the lung, the overall body radiation is going to be undetectable. That particle will be present and will lose half of its strength in 4,000 years. So that extensionally that particle is going to be stuck in your lung for the rest of your life, irradiating cells right around that particle! So the chances of lung cancer are extremely good. It is the most dangerous poison that mankind has ever faced! There is no question that there is a connection from those areas. Also, that milk from St. George did and still does go to Las Vegas.

INTERVIEWER: What might the effects have been on the people in this tract that were exposed?

NARRATOR: If there were concentrations of let's say iodine-131 in milk it would go to the thyroid.

A careful study by the University of Utah has recently concluded that the heaviest cancer and leukemia mortalities in the hardest hit areas of Utah have doubled and tripled. Contaminated milk from local production is believed responsible.

Las Vegas, just south of the test area was spared the direct effects of the fallout, but the recent news that the Las Vegans were the main consumers of the milk from those same hard-hit Utah areas could shake this desert valley.

THERE IS NO DANGER! That was the rallying cry of the Atomic Energy Commission throughout this period of Nevada's history. The public relations assurances and government films and announcements left a far different impression on some native Nevadans.

VICTIM NO. 6: The air would be just blue and sometimes purple. I happened to be going down to Alamo and about four miles this side, a deputy sheriff was there and cars were piled up. When I asked what it was all about, he said they had orders to wash all the cars coming through the area. I said, what about us breathing all this stuff? The air was purple, almost black. He shrugged his shoulders and left very angrily. They were washing cars for Vegas and here we were breathing it! [The fact was, Vegas was south upwind and the only threat came from the irradiated cars!]

VICTIM NO. 7: When all the radiation was being dropped in this country, right around this area, after we got our geiger counters, we would check them with the AEC. They would come in and we would say, "It's kinda hot today," and they would say, "No, there's no radiation today." Then we would get out our counters and check them against theirs and they would look kinda sheepish and take off. This sort of stuff really burns me up! [More truth than poetry in that statement.] Because they thought we were just dumb. Maybe we were for a while, until we were educated to the facts and what was going on. At one time they had AEC guys around here and we felt they were protecting us and we weren't guinea pigs anymore. There were many AEC trucks around watching us very carefully. We became very good friends with them because we weren't guinea pigs anymore, and they were protecting us. They claimed there were buses here ready to evacuate us. It got so hot, they left before anything actually happened. . . . the drivers weren't about to stay and be in it. There were some funny things happening.

NARRATOR: The reliability of government fallout data is now being seriously questioned. Some of the observed effects could not have occurred unless the fallout was ten or more times as extensive as these official maps indicate.

VICTIM NO. 8: I think that the criteria that they were operating under in those days was much more liberal than what they are using nowadays, indeed, when one considers that those were atmospheric tests. A burst occurred above the surface, on top of a tower, or dropped from a balloon, or dropped from an airplane, or shot from a cannon. Then it went off in the atmosphere and the familiar mushroom cloud formed, it went wherever the wind conditions carried it. There was an extensive program at that time to inform people about it.

They built contour maps to show us where the fallout went and they told us how they did it. They drove their cars down the road and hopped out and walked twenty-five feet to one side of the road and twenty-five feet to the other side and took the readings in that way. Someone indicated that someone should have checked the odometers on the vehicles. There was a 100 percent error when you tried a corroborative testing, when you used an independent contractor! That was because the odometers on the automobiles hadn't been checked. There is a lot of doubt about those figures.

INTERVIEWER: The government now admits that a reassessment of fallout data may be due. [All such maps are merely *estimated* and are often made by computer, *before* arrival of a radioactive cloud.] In cooperation with the states, the government is now undertaking a reappraisal of the fallout throughout the region. Using modern techniques and previously classified material, an updated data file has been promised by the Department of Energy. When you have this complete file, then you will use the data that you have for what we call a dose reassessment, to reconstruct what happened, to check data, the raw data, the other data, etc., to reconstruct the situation, the amount of fallout that was in a given area, where it went, how long it took to get there, what the meteorological, climatological situations were at the time, to look at the people that were there, to determine what kind of a dose they received, the exposure they received, both external and internal, and try to make an assessment of possible consequences from the radiation they received.

VICTIM NO. 7: Well they say it's just like an X ray, which only takes a couple of seconds or a second, right? How about us when we were under that stuff all day long.

NARRATOR: Were the community residents of the fifties and sixties used as guinea pigs? Of course we've heard this, and the answer is categorically no! These people were not used as guinea pigs in the sense that they were chosen to be exposed just to see what would happen to them. As a matter of fact, I cannot conceive President Truman or Eisenhower or Kennedy or Johnson or Nixon or any President of the United States, in any way, saying, that we would use a segment of our population, wherever, for guinea pigs.

INTERVIEWER: How do you feel, Mrs. Sharp, about all this now that it's all over?

VICTIM NO. 1: Well I guess it's all right now. [How little did they realize that that was only the beginning.]

THAT AMAZING AIR AROUND US

So that you will better understand what you have just read and what you are going to read, the following authoritative information comes from the Lawrence Livermore Laboratory Report. This agency is a unit of the ERDA (Energy Research and Development Administration). I like to make waves, but, *honest,* I didn't make the one-thousand footers.

The air ocean above us is in three layers. From the ground up to about 20,000 feet is the biosphere. This layer supports all growth and life as we know it. It then gives way to the troposphere where almost all weather happens. This layer is about five miles thick at the poles and ten miles thick at the equator. In the middle latitudes it tends to fluctuate from season to season, and even from day to day. The annual change can be from five to ten miles, while daily changes of a mile or more are common, particularly in winter or spring. These changes are like giant waves, thousands of feet high, on the top surface of the troposphere. There is, however, very little mixing of tropospheric air with that of the stratosphere above it.

From the top of the biosphere, the air grows increasingly colder with altitude. Most particulates—smoke, dust, volcanic ash, or radioactive debris—tend to fall back to the earth in rain or snow, within a month, from the troposphere. (Of course, they will circle the earth about every fourteen days. Pollutants placed in the stratosphere fall to earth in spurts over periods of months to years. Actually, they seed the clouds, causing heavier than usual precipitation, sometimes amounting to cloudburst proportions or severe blizzards, with extreme cold brought down with it.) The report states:

> The stratosphere overlies the troposphere, to a height of about thirty miles and is normally a very stable region, except for very severe thunderstorms whose tops occasionally climb above the troposphere. In spring and summer, there are no storms. Any particulates that get up into this area tend to stay there a *long* time. When a rare storm forces its way up into the stratosphere, it mixes and drags down material that would otherwise have remained there until pulled down by gravity

> or one of those giant waves. There are also local exchanges between tropospheric and stratospheric air above strong storm fronts.*

The coast range tends to baffle these storms somewhat. Frequently they spread over Canada and down through the central states as a nor'wester. Sometimes a large front splits and runs down the coast and over the coast range. It then bounces over the cascades and wham—it slams into the stone wall of the Continental Divide, the Rocky Mountains. Often it is diverted by them to the south, across the southern states, then up the east coast and out over the Atlantic. Europe can have it. This U-shape pattern is common on weather and temperature maps. See Fig. 1.

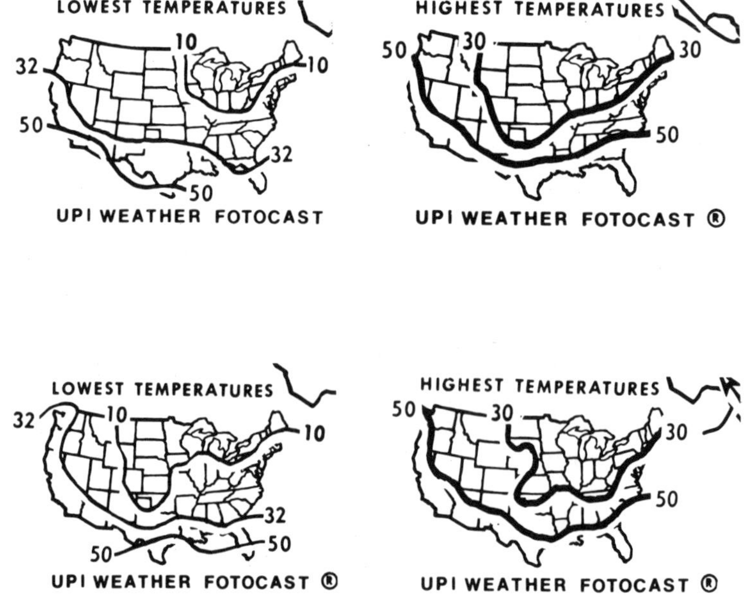

FIGURE 1

UPI weather maps. *(Used with permission)*

*These come roaring in from the Gulf of Alaska, where most of our weather is made. The Lawrence Livermore Laboratory Report ends at this point.

Well, so much for the air ocean above us. Now let's see what Mother Nature can do to this enormous ocean of air. Can we learn to differentiate between *Acts of God* and *Mother Nature?* The insurance companies are doing a lousy job of it!

ACTS OF MOTHER NATURE

Krakatoa Blows

The volcano Krakatoa blew her stack the first known time in 1680 and again in 1780, which was a mild one compared to the one of 1883. It all began on a small island in the Sundra Straights, about halfway between Java and Sumatra, in the South Pacific. The island was about five miles long by three miles wide, with two peaks. The largest peak was 2,700 feet high. The island was uninhabited at that time. Now let's go back 200 years, to the second eruption of 1780. At about 10:00 A.M., a darkness covered many parts of the United States so completely that the chickens returned to their roosts. Lamps and lanterns were required to see or read. It was not an eclipse. That night, what should have been a full moon was completely dark until almost dawn. When it began to reappear, much as the sun burns through fog, it was blood red! This was an act of Mother Nature, though it was construed by many as a sign from God.

The volcano had been dormant ever since the 1780 eruption. Passing ships observed in May of 1883 considerable volcanic activity. On August 27 of that year the volcano exploded (not an eruption). No pictures were available for that one! Most of the mountain was blown away. The shape of the island, as well as those close by, was radically changed.

Gigantic Tusnami (tidal) waves, generated by earthquakes, estimated to be 100 feet high formed. Since then, we have found out that the inundation height is five times the wave height! So don't go any closer to sea level than 500 feet to watch for a Tusnami! Watchers lined the small bluffs in Hawaii one time, watching for a giant wave and they, thirty-eight of them, I believe, were all swept away! The waves from Krakatoa swept over the shores of the nearby islands causing an estimated 50,000 casualties. Two new islands were formed where the day before there had been 180 to 250 feet of water! The sound of the explosion was heard hundreds of miles away. Volcanic dust circled the earth for years. Red sunsets were visible all over the world. The heavier rocks and boulders were actually, as we know now, thrown into orbit, to return years later as meteor showers. A falling of the stars from heaven, according to certain religious people of that time. We know now that the stars did not fall from heaven as a sign of the beginning of the end of the world, according to the Bible. It may never end! More about that later.

In those days little was known about the atmosphere, or the oceans, or the weather. The God-given makeup of our world, according to history, was much the same then as it is now, except for the quirks of Mother Nature. The higher you climbed, the thinner the air became, until only heaven was above.

To explain this belief in signs: The Disciples of Jesus asked him to tell them how the end of the world would begin. Jesus said, "The sun will not shine and the moon will not give its light [which has happened several times since], and the stars will fall from Heaven. . . . Then, they shall see the Son of Man coming on the clouds of Heaven!" Matthew 24: 29-36; Isaiah 13: 9-13; Joel 2: 31. This very thing happened during World War I. Perhaps Jesus, knowing they would not understand, said it thus for the benefit of those people who would be living hundreds or thousands of years in the future, knowing that they would understand. That it would *never* happen as far as total destruction of the earth is concerned! A new earth and a new heaven is promised after destruction by fire from the skies! As people of our time would say, "It'll be a cold day in hell, when the stars fall from the sky."

There was a great meteorite shower, which consisted not of stars but, we know now, of orbited rock and boulders from volcanoes. Who knew anything about orbits in those days? It was construed to be the stars falling from heaven at that time. Though the next night the stars were seen to be still in the heavens! Jesus did say, however, "As for that day and hour, nobody knows it, neither the angels in heaven, nor the Son, no one but the Father only." We know now that nothing can be totally destroyed! But the form or substance can be changed, reverting to smoke, ashes, gas, or whatever. The first so-called destruction was by flood. The second time around will be by fire from the skies, and then will come the "New Heaven and

New Earth"! So behave yourself, tread the straight and narrow. It is later than you think.

Worldwide communication was nonexistent until recently. Apparently, a fault exists in the earth's crust under old Krakatoa. From recent events it would appear that every hundred years, give or take two or three years, 1680, 1780, 1883, there is an eruption. Will it blow again in 1982 or '83?

Krakatoa's 1883 volcanic ash brought on the worst worldwide winter and the greatest blizzard in the history of the east coast; ash blanketed the earth from the heat of the sun. In that year, gravity finally got a good hold on that ash and debris, plus ash from other volcanoes that were very active in those times, and pulled it down all over the world.

Nothing was known about cloud-seeding in those days. We now know that is just what happened. This doubled the severity of the winter over whatever was normal, wherever it happened. The weather went to pot all over the world!

Mr. Patrick Hughes of the U.S. Weather Bureau reported the weather history for the Northeast at that time. The forecast for Sunday, March 11, 1888, was cloudy, light rain, and clearing on Monday. Instead, the drizzle turned to snow. Evidently an upper air wave had crashed on the East Coast, bringing extreme cold and zero temperatures, with 70 mph winds. It piled drifts forty to fifty feet high and buried entire villages. For days the northeastern United States was paralyzed. Trains were stalled. Even the largest cities ground to a halt! Tunnels were dug and shored up, as in mines. Over 200 people died in New York City alone, with over 400 for the area total! High winds, deep snow, and low temperatures had occurred many times, but never all at once. People trapped in New York slept on the floor of armories or wandered from one hotel to another, begging for a place to rest. Babies born during the blizzard were named for it. After forty-eight hours, the full scale blizzard subsided. To this day, no one apparently has put two and two together to determine the cause. Probably no one cared up to now, or perhaps it took the alert mind of a researcher to relate the facts lying dormant all these years. Evidently, when old Krakatoa blew her stack in 1888, it started the whole thing. With our knowledge now of the air above us, it has been accepted as the cause of the bad weather all over the world.

Other, more ancient, safety valves have contributed their share of meteorites in orbit and volcanic ash.

FIGURE 2

A New York City street during the Blizzard of 1888. *(Wide World photo)*

FIGURE 2a

This photo shows the depth of the snow that clogged the sidewalks.

Vesuvius

Vesuvius, a better-known earthly safety valve than Krakatoa, has also had a part in affecting our global weather. The first eruption occurred in A.D. 63. Not until the second in A.D. 79, when the ancient cities of Herculaneum and Pompeii were destroyed, was the event recorded. Hot poisonous gases instantly cut off the oxygen supply after burning every thing that would burn. The first eruption in A.D. 63, apparently left no one to record the destruction! Whole cities were buried by volcanic ash. It is thought that the great earthquake of Nicomedia, in the year A.D. 358, 279 years after the A.D. 79 eruption, relieved much of the stress that would otherwise have caused a devastating blowout. Just before the quake, a mysterious phenomenon developed. Darkness, very dense, developed in midday, lasting from two to three hours; stars were visible. Two years later, in A.D. 360, all of the Roman Empire was in total darkness from dawn until noon. Stars were again visible during that time. The possibility of an eclipse was ruled out because of its duration. (A possible planetary conjunction?) As light began to return, the sun first appeared as a crescent, then half, until it was gradually restored to its full size. That may not have been an eclipse as we know it, but you can bet that one of the nine planets, large enough to cover us from the sun, got in the way. Venus with its 7,700 miles in diameter could do the job. Our good old earth with a diameter of 7,918 miles, is only 218 miles larger and would be pretty well blacked out!

After 957 years, Vesuvius erupted in A.D. 1036, with many violent eruptions occurring off and on up to the year 1631, 595 years later, when, even with the numerous pop-offs, it really blew and caused the death of over 18,000 people! More ash in the sky and more boulders in orbit! One hundred and ninety-one years later, another eruption occurred, in 1822. Then sixteen years later, in 1838; another in 1872, thirty-four years later, a very violent one. Eruptions occurred eight years later, in 1880; five years later, in 1885; twenty-one years later, in 1906; twenty-three years later, in 1929, and fifteen years later, in 1944. The world was much too engrossed in war to pay much attention to the last one. A lot more boulders were put in orbit for more meteorites and volcanic ash for cloud-seeding!

Three months after Krakatoa blew, in November, 1883, a great meteorite shower occurred. Huge fireballs were common;

shooting stars were everywhere! Were they from Vesuvius eruptions of 1631 or 1880, or from Hawaii's Kilauea, Mauna Loa, Kohala, or two older ones now extinct? Mauna Loa, measured from the ocean floor to the summit, is the greatest volcanic mountain on earth, with 18,000 feet from the floor of the ocean to sea level, then on up to 32,000 feet to the summit. The Hawaiian volcano Kilauea, 10,000 feet lower than Mauna Loa, is still active; measured from the ocean floor it is 22,000 feet, bottom to top, a gigantic volcano. The islands are of volcanic origin and still growing. One eruption of Mauna Loa (they don't always blow from the top) opened up on the side and poured out a lava flow that was seven miles wide and ran forty-four miles to the sea. This river of lava continued to flow for twelve months! It added many square miles to the big island of Hawaii.

Many dark days could have been caused by these mountains before the islands were even populated or discovered. Mother Nature was just blowing her stack! Now let's get off the stack and down to earth. Many people have been in hurricanes and typhoons and lived. Many have died because of them. The worst typhoon ever generated in the Pacific dissipated. Man's foul-ups seem to last forever and ever. The Thomas Typhoon, named for the ship shown in Fig. 3, was a ripsnorting windstorm! The *Thomas* was the only large ship to come through it alive. Hundreds of small vessels were never heard from again!

The Thomas Typhoon

The *General Thomas* was an army transport ship of 7,684 tons, Gr. Wt., twin screws, built in Belfast, Ireland, in 1894, and damn well built she was. She was purchased in 1899 from the British, and her name was changed from *Minnewaska* to *General Thomas.* Rebuilt in 1911, she was on her ninety-fifth round trip to the Philippines in 1926. I was aboard. At that time, she had covered about a million and a quarter miles of ocean, making six round trips safely a year. Her slowest time in twenty-four hours was 197 miles; the fastest time was 325 miles. A coal burner, she used seventy tons a day. She coaled up at San Francisco, Honolulu, Manila, and Nagasaki.

Very picturesque were those small, female, Japanese coal stokers. Like a string of ants, they trod up the gangplank into the bunkers, where they dumped their basket of coal. Then on

FIGURE 3

The USS *General Thomas. (Courtesy Oregon Historical Society)*

to another side opening and down another gangplank for another basket of coal. Such sturdy legs you never saw before. A regular round robin, a human conveyor system. Each woman carried a large basket of coal up the plank and frequently came out the other door with a tee shirt or a pair of shoes in her basket. This was, no doubt, a gratuity for a favor dispensed.

A passenger gave this eyewitness account of the storm at sea:

> The typhoon lasted from August 13 to August 21, 1917. The *Thomas* was on her way from Manila to Nagasaki for coal. The typhoon caught her off the coast of Formosa. The fury of the elements as experienced by the *Thomas* for the next two or three days was such that it was considered a miracle that she wasn't a complete wreck.
>
> By Friday, August 17, we were well into the typhoon. The wind was violent and the sea both magnificent and awful. In a typhoon the wind blows the surface of the water off into flying clouds of foam and spray. For all the world, it looked

like a snowstorm at sea. The water could not be kept out. It came in through the ports and through seams in the cabin roofs. On the top deck, stateroom doors were forced and the rooms flooded. Beds were wet and cabin floors awash. The ship hove to, and nosed into the wind. We made only eight miles in fifteen hours under steam. No one knew how far from our course we had drifted.

From experience, mariners know that the way to get out of a typhoon is to keep the wind on the starboard bow and side, and gradually work your way out with the help of the wind. This the ship's master did not dare to do, as from the position he thought we were in, chances were good that we would be cast on one of the many islands. He therefore elected to keep the wind on the port side and run through the typhoon. Suddenly, there came a lull with no air stirring, a calm sea, the barometer having fallen with nearly a perpendicular curve. The pressure on the eardrums was intense. It was very plain that we were in the very center of the typhoon, with our most severe trial with the elements just ahead. Some time around 4:00 A.M. the wind rose. It soon became terrific, well over 100 mph. The fight was on. The first officer estimated it at 125 to 130 mph, by far the worst we had experienced. Now ensued a time of desperate peril that even a landlubber could appreciate. Many a prayer went up that night from lips that had not uttered one since the days of "Now I lay me down to sleep," at a mother's knee.

The awful force of the wind, the violent pitching and rolling, the absolute darkness outside, the puny efforts of the big, powerful ship and its apparent helplessness added to the fact that all were aware we were in an unknown position. This, in a sea that in many places is uncharted and dotted with coral reefs and islands. All were convinced that we were in the presence of the Great Adventure, and that only the guidance of the Divine hand could save us. The immediate scenes were enough to inspire terror. The big steam whistle could barely be heard above the wind when it blew to call the deck hands to clear away the wreckage caused by the fall of the top of the front mast, with the wireless antenna. The galley was right outside my door.

The pots and pans, dishes and glassware were thrown in continuous din and confusion from one side of the galley to the other. In staterooms, trunks, water bowls, and loose articles were flying around. It was dangerous to be about. The thick glass of a porthole was broken to bits by a wave, and a woman nearby was so cut by glass that she required many stitches for her wounds. Many bones were broken and many persons severely bruised. In the troop decks many bunks were broken and fell to the floor. Many soldiers suffered broken legs, arms, and collar bones. This violence kept up most of the day of Sunday, August, 19. We were destined

> for the third successive sleepless night. About 11:00 P.M., the ship struck something lightly twice and in turn was struck on the starboard side by two big waves. The vessel remained rigid for several minutes, estimates were from five to twenty-five minutes. During this time a soldier was swept overboard; his cries for help were heard by passengers as the wave swept him past. A life preserver was thrown toward him in the darkness, all that could be done for the poor fellow.
>
> About this time the lookout ahead saw a steep black shape close at hand, but thought it was a low-hanging cloud and said nothing, or at least before he could, the first officer saw it from the bridge and immediately signaled to the engine room for emergency. We were merely drifting and the engines were immediately reversed. The ship drew away carefully. The wireless was put out of commission; when the top of the mast went Saturday night, the antenna went with it. The latter was roughly secured again to the main mast, but the violent wind had driven the rain through the cabin and short circuited the sending apparatus.

I had learned the situation by this time and was convinced in my own mind that it was our last hour. Since then I discovered that that belief was universal among all but a few who were in ignorance of the conditions. During the rest of the night, we tossed about the ocean, by some chance avoiding the shoals and reefs of the Ryukyu group in the vicinity of Irimote Island, as was later determined.

The typhoon slowly abated, and it was not known until 2:00 P.M. August 20 what the ship's position was. She was 100 miles northeast of where she was estimated to be. Her position was unknown for over two days after the storm subsided.

Now, there was an act of Mother Nature, unaided by any atom bomb—a true ocean-born typhoon. Nature can come up with some terrible storms on her own without any help from mankind and atomic weapons to upset her apple cart!

The Dark Days

In the year 409, darkness was so complete during some days that stars were visible. About the year 536, the sun was blocked out so that little of its light was seen for fourteen months! (It just had to be planetary conjunction.) In 567, darkness prevailed from 3:00 P.M. until night fall, so complete that nothing was visible. In the year 626, half of the sun's disc was blocked out for over six months! In 733, 107 years later, it happened again. People were terrified. In Portugal, darkness persisted for two months with the sun returning to dimness, just before the heavens were opened by fissures and huge bolts of lightning; suddenly the sun was as bright as ever, and all seemed to be well with God's world of that time.

However, on September 21, 1091, the sun was blocked out for three hours. On February 28, 1206, complete darkness prevailed for six hours during the day. In the year 1241, the stars were visible at 3:00 P.M. In 1547, on April 23 and 25, again darkness prevailed and stars became visible during the day. Now, skipping past Krakatoa's second blow of 1780 to 1790, ten years later in New England, on May 19, it began to grow dark around 10:00 A.M. continuing in intensity so that lanterns were required by noon. By chore time in the evening, it was possible to see the lanterns but they did not throw off enough light to see by! The black air was thick and heavy and unable to carry light rays! Birds and poultry went to roost at 11:00 A.M. on that day. When Mount St. Helens blew, cities to the east had that same problem!

Now some of this we will have to blame on Mother Nature, as we know she can really foul up the weather. As for other reasons that these things occurred, we are only now beginning to formulate a theory now and then.

Two scientists have forecast a cosmic event of incredible significance for 1982. According to Stephen Plagemann, researcher at NASA's Goddard Space Center, and John Gribbin, physical science editor of the British magazine *Nature,* an alignment of all nine planets in our solar system will take place on one side of the sun. They predict that dire things will happen in the solar system because of this phenomenon. They also state that planetary alignment happens about every 179 years, and that this will be the first time since creation that a *perfect* alignment will take place.

Now just a cotton-pickin' minute! I can go for a sloppy lineup at various times, which would account for many of the *dark days* of history and other unaccountable happenings—as I look at an orbital map of the nine planets, it seems that if Mercury or Venus were properly lined up between Earth and the sun, some of our *dark days* could very easily have been blamed on them—but since the planets are held in their orbits by the sun, I fail to see how perfect alignment can hurt Mercury, though a prolonged eclipse of Venus, Earth, and Mars might cause quite a disturbance on those planets. However, because gravity is outgoing from the sun I doubt that the others in the far-out orbits would be greatly affected. Their gravitational pull from the sun would be lessened if anything. Their orbits might even be enlarged and then pulled back when the fracas was over. Also, due to their size, it is doubtful that the small shadow cast on them from the closer-in planets would cut off enough of the sun's gravity to even make them wobble! Plagemann and Gribbin's book, *The Jupiter Effect,* should make interesting reading, if you are interested in earthquakes and the possibility of predicting them.

James Hansen of the Goddard Institute for Space Studies says, "*The density* of the atmosphere (measured near the surface by two of the Venus probes, as 90.5 and 91.5 bars) would *bend light so sharply* from a straight path that an observer gazing down from orbit *might see no planet at all.* Looking down on even a haze-free Venus, one infers, might be like looking at empty sky!

This atmosphere density on our own Earth could account for the *dark days,* due to volcanic ash, and so on, bending the sun's rays or reflecting them right out into space until such time as gravity cleared the upper air enough to allow the penetration of light! This might also account for the black holes in space! It would also affect our weather—and how! Whether these actions are those of God, or laws of the universe, take your choice. Mother Nature's laws must be strictly confined to Earth, and the U.S. Department of Commerce has collected some very interesting data on weather as presented by Mother Nature.

Highest Temperature Extremes in the United States

The average normal temperature for the United States is 78.2 degrees. Amazing temperature rises of forty to fifty degrees, in a few minutes occasionally may be brought about by the chinnook winds.

Greenland Ranch, California, with 134 degrees, on July 10, 1913, holds the record for the highest temperature ever reported in the United States. This station was located in barren Death Valley, 178 feet below sea level. Death Valley is about 140 miles long and from 4 to 16 miles wide and runs north and south in southwestern California. Much of the valley is below sea level and is flanked by towering mountain ranges, with Mt. Whitney, the highest landmark in the forty-eight conterminous states, rising to 14,495 feet above sea level. Fewer than 100 miles west, Death Valley has the hottest summers in the Western Hemisphere, and is the only known place in the United States where nighttime temperatures sometimes remain above 100 degrees. Other notable temperature rises (Fahrenheit) are listed below.

TABLE 1

Highest Temperature Extremes in the U.S.

Time period	*Degree rise*	*Location*
12 hrs.	83°	Granville, North Dakota, February 21, 1918 from −1° to 50°, from early morning to late afternoon
15 min.	42°	Fort Assiniboine, Montana, January 19, 1892, from −5° to 37°
7 min.	34°	Kipp, Montana, December 1, 1896. The observer also reported that a total rise of 80°, occurred in a few hours, and that 30 inches of snow disappeared in half a day.
2 min.	49°	Spearfish, South Dakota, January 22, 1943, from −4° at 7:30 A.M. to 45° at 7:32 A.M.

Lowest Temperature Extremes in the United States

The lowest temperature on record in the United States, −79.8 degrees, was recorded at Prospect Creek Camp in the Endicott Mountains of Northern Alaska (latitude 66°48′ north, longitude 150°40′ west) on January 23, 1971. The lowest ever recorded in the conterminous forty-eight states, −69.7 degrees, was observed at Rogers Pass, in Lewis and Clark County, Montana, on January 20, 1954. Rogers Pass is in mountainous and heavily forested terrain about half a mile east of and 140 feet below the summit of the Continental Divide.

In the continental areas of the temperate zone, temperature falls of 40 to 50 degrees in a few hours caused by advection are not uncommon. Advection is heat transfer by the horizontal motion of air. Sometimes these large temperature changes are followed by additional heat loss through radiation, resulting in remarkable temperature falls. Those giant waves on the top of the troposphere are boosted over the Rockies and come crashing down the other side with the extreme cold of the upper air, sliding along the ground of Montana and the Dakotas under the comparative warmer air above before curving up again to where it belongs, taking with it whatever heat is close to the surface of the earth by radiation transfer. The table below lists some big temperature (Fahrenheit) drops that have taken place.

TABLE 2

Lowest Temperature Extremes in the U.S.

Time period	*Degree drops*	*Location*
24 hrs.	100°	Browning, Montana, January 23, 24, 1916, from 44° to −56°
12 hrs.	84°	Fairfield, Montana, December 24, 1924, from 63° at noon to −21°, at midnight
2 hrs.	62°	Rapid City, South Dakota, January 12, 1911, from 49° at 6:00 A.M. to −13° at 8:00 A.M.
27 min.	58°	Spearfish, South Dakota, January 22, 1943, from 54° at 9:00 A.M. to −4° at 9:27 A.M.
15 min.	47°	Rapid City, South Dakota, January 10, 1911, from 55° at 7:00 A.M. to 8° at 7:15 A.M.

As you can see, Rapid City, South Dakota, really had its ups and downs in January, 1911. Spearfish, South Dakota, had it even tougher! All this came about without the aid of atmospheric bombs. Though very cold weather is not uncommon, this was unusual for the area.

Before we get into the heavy stuff of atomic weapons and who "they" really are, let's look at the so-called miracles of the Bible.

FIGURE 4

Mount St. Helens erupted on May 18, 1980. Plumes of ash and ice blew many times higher than the 9,677 foot mountain itself. *(Photo by Don Wilson, used with permission of Ross Hamilton)*

The Mount St. Helens Eruption

The volcanic ash from Mount St. Helens is as fine as talcum powder and seems to be a blessing in a way. There have been no yellow jackets to annoy picnickers since the eruption. Then again, it smothered hundreds of hives of bees. That was not good. It has not been determined whether the bees died from radiation or from being smothered.

According to a study by the Batelle Pacific Laboratories in Richland, Washington, rather high concentrations of several radioactive elements were found at Spokane, 227 miles from the volcano, after the May 18 eruption. Was it scavenged from the stratosphere? The ash cloud was blown up to well over 70,000 feet! The radon gas released radiation greater than TMI.

The study by Dr. Richard Perkins of the Batelle Laboratory concerns an extensive analysis of volcanic ash that fell at various points across the state of Washington. It reports that isotopes of radium-226, potassium-40, thorium-232, cononium-210, and lead-214 were found in the ash! This went around the world and is still coming down in rainouts, snowouts, and by gravity.

The birds were all in trees or shrubs. None were on the ground. Their wings were spread, mouths open, and tongues out in an effort to breathe. Animals everywhere were having a terrifying time. People weren't doing so well either.

ACTS OF GOD AND MIRACLES IN AND OUT OF THE BIBLE

The Bible is full of miracles, for example, the burning bush that was not consumed (Exodus 3:1-6)! An angel of Yahweh (*God*) appeared before Moses in the shape of a flame of fire, coming from the middle of a bush. Moses looked at the burning bush that was blazing but not being burned up. "And Moses said, 'I must go and look at this strange sight, and see why the bush is not burned.' " God called to him from the bush and said, "Come no nearer, take off your shoes, for the place on which you stand is holy ground." For the rest of the miracle, go to the Bible; you will find it very interesting. The next miracle is the most marvelous of all.

The Sacrifice on Mount Carmel

Ahab called all of Israel together and assembled the prophets on Mount Carmel. The prophet Elijah stepped out in front of all the people. "How long," he said, "do you mean to hobble on first one leg and then the other? If Yahweh is God, follow him." The people did not answer. Elijah said to them, "I alone, am left as a prophet of Yahweh, while the prophets of Baal are four hundred and half a hundred. Let two bulls be given up, let them choose one for themselves to sacrifice. Dismember it and lay it on the wood but do not set fire to it. I in turn will prepare the other bull, but not set fire to it. You prophets of Baal must call on the name of your God, and I shall call on the name of mine. The God who answers with fire, is God indeed." The people all answered, "Agreed."

The prophets of Baal called from morning to midday on the name of Baal, but there was no answer. The time of sacrifice would soon be over. Elijah ridiculed them. He said, "The prophets of Baal must call louder, for your God is asleep, or busy or gone away." The prophets of Baal slashed themselves with swords and sprears, as was their custom, but no answer, no fire came, and they gave up!

Elijah called the people to come closer to him. He repaired the altar of Yahweh, which had been destroyed, with twelve stones, representing the twelve tribes of Israel. He dressed the bull and laid it on the wood. Then he ordered four large urns of water poured all over the carcass and the wood. He had this done three times, until a trench he had dug around the altar was filled. Everything was saturated. Elijah then stepped forward and called on Yahweh, the God of Israel. Fire descended from heaven and consumed the offering, the wood and water, and even the stones of the altar! The people fell on their faces and acknowledged that Yahweh is God (1 Kings 18:20-40)! Look at the account in your Bible. It is so much more interesting than my condensed version.

When God took Elijah into heaven, Elijah was replaced with Elisha. As they stood by the river Jordan, Elijah took off his cloak, rolled it up, and struck the water. It divided to the left and right. They crossed the river dry shod. On the other side, Elijah was taken to heaven by a chariot of fire pulled by horses of fire. His cloak fell to the ground. Elisha picked it up, tore his clothing off, and rolled the cloak as Elijah did, striking the water and calling on God. He recrossed the river as the water parted. The brotherhood of prophets watching in the distance saw this (2 Kings 2:1-18).

Elisha performed two more miracles. One good; he purified water. Later he cursed a group of boys who made fun of him: two she-bears came out of the woods and mauled forty-two of them. (I wonder if that would work on today's motorcycle gangs? For a prophet, perhaps, it would!) Now for God's miracle for Moses.

The Plague of Darkness

The Lord said to Moses, "Stretch out your hand toward the sky so that darkness will spread over Egypt, darkness that can be felt!" So Moses stretched out his hand toward the sky, and total darkness covered all of Egypt for three days. Yet the Israelites had light in the places where they lived! See Exodus 10:21-23. Then came the parting of the waters so the Israelites could escape the Egyptians.

Another darkness occurred at Christ's Crucifixion, lasting from noon to the ninth hour!

Other miracles in the Bible too numerous to mention attest the power of faith. In our time Billy Graham prayed over a box full of names of sick and dying people, and they were cured. Saints, long gone, still exert power for good all over the world. Healing and answers to prayers continue.

Christ, in two paintings by Peter Hurkos, wept, one from the wounds on his hands, and the other from the Crown of Thorns. A painting of Moses began weeping tears that turned to blood. Again, two years later, a plastic-enclosed portrait of Christ was seen seeping blood from the right eye. The blood came through the plastic and lasted about fifteen minutes. A reporter, Al Giben, from the local paper of Roswell, New Mexico, saw the painting and called in the chief medical technician from the hospital, who confirmed that it was blood. To see the blood flow from a plastic picture of Jesus, to watch it flow, knowing there could be no other explanation but supernatural power, was a terrific shock.

"The Virgin Mary Appears in Egypt: Performs Miracles!"

This event took place in April, 1968; the story appeared as a featured article by Craïg Chilton in the *Greater Oregon* newspaper, December 20, 1968, under the headline quoted above.

> As Christmas approaches, a new, joyous event of tremendous religious importance is happening near Jesus' birthplace. However, for some unknown reason, America has not yet been informed. [As far as is known, America was never informed].
>
> Eight months ago, two car mechanics who were working after midnight thought they saw a potential case of suicide atop the church across the street. They immediately notified the police and a priest, who were quick to arrive at the scene.
>
> This was the beginning of the most wondrous Christian event since the time of Christ. When the authorities arrived, the figure dressed in white above the church was recognized as being a miraculous apparition of the Virgin Mary!
>
> Since April 2, 1968, the Virgin has made many such appearances over the Coptic Orthodox Church of the Virgin Mary in Zeitoun, a suburb of Cairo, Egypt. During the month of April, she appeared nearly every morning, including every day of Holy Week. From then until late September, she appeared on the average of two to three times per week. Since that time, she began appearing regularly on Sunday mornings exclusively. . . .
>
> This event has caused such interest in Europe and Africa that more than 20,000 pilgrims a day were flocking to the site. The church is believed to be located on a stopping point for the Holy Family on their flight to Egypt.
>
> In a world that largely does not believe in miracles, the events have proven particularly significant, since a number of healings have occurred in conjunction with the visions, almost from the beginning.
>
> Mr. Manfouz Doss, board chairman of the Los Angeles Foreign Press Association, regularly receives the Coptic newspaper *Watani* from Cairo, which has provided much current information.
>
> Mr. Doss reports that cases of actual healing are published weekly in *Watani,* but only after they have been completely documented and verified by competent physicians and specialists who know the condition of those healed, both before and after the miracle.
>
> One example of miraculous healing is taken from the October 27 issue of *Watani.* On the previous Sunday, October 20, the Virgin Mary appeared twice. The second time, her appearance was much clearer than the first, as she appeared in a glowing light. At that particular time many healings occurred. Others followed after the people went home.

"Mrs. Farka Gobrial, wife of a Cairo jeweler, had a serious heart condition, as well as arthritis and a serious disc condition in the back. Her problem was considerably worsened by the fact that the medicine for her back aggravated her heart, while her heart medicine aggravated her back.

"She went to the Coptic church and prayed fervently that she be healed. Suddenly, she literally felt a hand going over her back—it felt cold, like a chilled hand. She turned to see who was doing this but only saw a woman standing a few feet away busy with her prayers. It was then that she noticed she could stand erect. Her back pain was gone. Later, she discovered her heart condition had been cured as well."

Many more people have been cured of disease, ranging from asthma to paralysis, to terminal cancer. Faith has not been a limitation. Egypt is a Moslem country, with about ten percent of the population belonging to the Coptic faith; one of the four original apostolic churches founded shortly after the time of Christ.

Many Moslems have been healed, as well as Christians, since the apparitions began. Brochures containing the full details were published over the years.

Strangely, this phenomenon [miracle], has been virtually ignored by most of the major news media in this country. It is lack of faith? [Or is it fear that a more or less Godless country might wake up and find that *God is not dead?*] Many opportunities have been made available in Egypt to get the details.

In June of this year [1968], representatives of 172 faiths, from American Presbyterian to Russian Orthodox, were present at the dedication of a new coptic cathedral in Cairo. Since the apparition had at that time been continuing for more than two months, undoubtedly they knew the details of these events before they went home.

Earlier, on May 4, 1968, an official statement by His Holiness Pope Kyrillos VI, Coptic Pope of Alexandria and Africa, was issued following intensive investigation by the Coptic Orthodox Church and other agencies, to a contingent of 150 reporters, journalists, and correspondents from around the world.

The story was released in Europe where it has created a sensation. It is only fitting that at this time of year, America learn of the modern Christmas Story—the rebirth and renewal of Christianity.

As yet, the Blessed Mother has given no messages, yet her presence in Egypt is a source of hope and inspiration to the world's teeming millions. America can be even more cheerful than usual this Christmas for many reasons: The coming peace talks may well end all wars; our new president will strive to bring us forward together in unity and restore law and order; and America now knows that God is not dead, but still cares for a sinful world.

The Bible Proves a Computer Correct

The following article was copied from the *Evening Star,* a Spencer, Indiana, newspaper.

> Did you know that the space program is busy proving that what has been called "Myth" in the Bible is true? Mr. Harold Hill, President of the Curtis Engine Company in Baltimore, Maryland, and a consultant in the space program, relates the following development.
>
> "I think one of the most amazing things that God has for us today happened recently to our astronauts and space scientists at Greenbelt, Maryland. They were checking the position of the sun, moon, and planets out in space, where they would be 100 years and 1,000 years from now. We have to know this so we don't send a satellite up and have it bump into something later on in its orbits. We have to lay out the orbits so the whole thing will not bog down. They ran the computer measurement back and forth over the centuries and it came to a halt. The computer stopped and put up a red signal, which meant that there was something wrong either with the information fed into it or the results as compared to the standards. They called the service department to check it out and were told, 'It's perfect.' The head of operations asked, 'What's wrong?'
>
> " 'Well, we have found there is a day missing in space in elapsed time.' They scratched their heads and tore their hair. There was no answer.
>
> "One religious fellow on the team said, "You know, one time when I was in Sunday School they talked about the sun standing still.' The others didn't believe him, but they didn't have any other answer, so they said, 'Show us.' He got a Bible and went back to the book of Joshua where they found a pretty ridiculous statement for anybody who has common sense!
>
> "There they found the Lord saying to Joshua, 'Fear them not, for I have delivered them into thine hand. There shall not a man of them stand before thee.' Joshua was concerned because he was surrounded by the enemy and if darkness fell, they would overpower them. So Joshua asked the Lord to make the sun stand still. That's right—'The sun stood still and the moon stayed . . . and hastened not to go down about a whole day.' The spacemen said, 'There is the missing day.' They checked the computers going back to the time it was written and found it was close, but not close enough. The elapsed time that was missing back in Joshua's day was 23 hours and 20 minutes—not a whole day. They read the Bible and there it was—'*about* a day.'

"These little words in the Bible are important. But they were still in trouble because if you cannot account for 40 minutes, you'll still be in trouble 1,000 years from now. Forty minutes had to be found because it can be multiplied many times over in orbits. This religious fellow also remembered somewhere in the Bible where it said the sun went backwards. The spacemen told him he was out of his mind. But they got out the good book and read these words in 2 Kings: 'Hezekiah, on his deathbed, was visited by the prophet Isaiah who told him he was not going to die. Hezekiah asked for a sign as proof. Isaiah said, "Do you want the sun to go ahead ten degrees?" Hezekiah said, "It's nothing for the sun to go ahead ten degrees, but let the shadow return backward ten degrees."

"Isaiah spoke to the Lord and the Lord brought the shadow ten degrees *backward!* Ten degrees is exactly 40 minutes. Twenty-three hours and 20 minutes in Joshua, plus 40 minutes in 2 Kings, make the missing 24 hours the space travelers had to log in the log book as being the missing day in the universe. Isn't that amazing? [References: Joshua 10: 8-14; 2 Kings 20: 1-11].

THE VORTEXES

A northwestern weather observer, comfortably situated in his easy chair at home, having slept through a TV show he really wanted to see, noticed a weather satellite photo on the screen. Having had a good nap, he was probably more alert than usual. He saw a vortex high up in the gulf of Alaska, gathering up clouds from the ocean cloud cover before it reached the coast. This huge pinwheel then flung the clouds far out in the Pacific again, past the Aleutian Islands. The vortex was moving counter-clockwise, as it should, which could make for a long dry spell all by itself. The unnatural part of the picture, however, was that just below this was another vortex spinning *clockwise,* also gathering up clouds from the ocean cloud cover off the coast of California and Bahai, spinning them off again far out in the Pacific. Now *this* is a *no-no* in the northern Hemisphere and against the laws of Mother Nature and God. *All* vortexes north of the equator are required to spin *counterclockwise!* Fig. 5 is a composite picture because a huge, radioactive debris cloud almost obliterated the swirls, for all practical purposes, by intervening between normal cloud cover and the satellite, practically exposing the picture. I have this photo, but it is not suitable for publication.

This event happened after an unusually dry summer. It wasn't

FIGURE 5

A composite picture showing the abnormal spin of the lower vortex.

helping things a bit. Streams were exceedingly low, some were drying up (see Fig. 6). The snow runoff was long gone. The northwest was running out of hydroelectric and nuclear power. Diesel and coal-fired generating plants were hastily started and put on the lines. Much power was purchased from other states. Something was radically wrong. Our weather observer decided to find out, if possible, what it was. You read about it in all the

FIGURE 6

A satellite photo of November 1, 1978, with grids, showing total absence of clouds, the cause of the drought in the northern hemisphere. *(Courtesy U.S. Meteorological Service, Portland, Oregon)*

papers. Snorkel divers were afraid to go into the rivers because the fish were too thick. Some were almost as big, some bigger than a man (see Fig. 7). Being interested in cause and effect, the weather observer decided to backtrack the weather satellite photos. These pictures taken from 23,000 miles up revealed that the vortex circus occurred shortly after September 26, 1976.

FIGURE 7

A 10-foot 9-inch sturgeon caught in the Columbia River. *(Used with permission of Don Holms and Joseph Bianco, editor of* Northwest *of the* Sunday Oregonian)

2

Who They Are and What They Did!

THE DAY OF THE SECOND SUNRISE

On July 16, 1945, a few minutes before dawn, the first atom bomb was detonated at the White Sands test site, near Alamogordo, New Mexico. This coincided with the graveyard crews going off shift in the various West Coast shipyards. Suddenly, a huge flare of white light appeared over the mountains in the east, like a rapidly rising sun! This predawn flare climbed like a rapidly growing mushroom; it remained suspended for a second or so, as though it intended to push on out into space but couldn't quite make it. Like a snowball, it suddenly melted from the top down, and it was gone as quickly as it came. Many of you no doubt saw it.

Seen from Oregon, Washington, and California, it aroused intense speculation that a huge meteor might have struck the earth or an ammunition magazine might have blown up. Minutes later the true dawn arrived, followed by the real sunrise. After the atom bomb was dropped on Japan, the media was informed that those who saw the false sunrise weeks before had actually witnessed the culmination of a great, secret development!

That fake, first sunrise would never set. Scientists and the military had found a new plaything. We are now enjoying the longest period of peace from big world wars the world has ever known. *No one dares* to use atomic weapons in war, as the boomerang effect is inevitable. *The good Lord must have arranged it that way!* Whatever is sent out or up will circle the earth in about fourteen days and land on the senders. The rest of the world will also be doused with that fire from the skies and *they won't like that.* If they can retaliate, you can be sure *they will.* Even playing around with it, testing, is deadly. To give a better picture of what atmospheric atom bombs can do—and have done—to the ocean of air above us, let's use an analogy.

THE ROCK IN THE POND

Suppose you take as large a rock as you can carry and drop it in a pond or slow-moving creek. It splashes water all over, making waves galore and a fountain of water, smashing ripples and currents, raising cain on the bottom. This will compare to the September 26, 1976, atmospheric test atomic bomb of 20/200 kilotons that the People's Republic of China sent up into the troposphere. This was equal to 200,000 tons of TNT! Now take a boulder twenty times as large, which of course you couldn't lift, and before the mud settles and things return to normal, drop it right in the same place. WOW! Not enough left for the fish to swim in. On November 17, 1976, the Chinese sent up a 4 megaton bomb, twenty times larger than the *little feller!* This one was equal to 4 million tons of TNT! It was sent up right smack dab into the center of the debris cloud from the first one of September 26 that was still circling the earth! *That* really batted out a radioactive debris cloud, blanketing the world! Jet streams were knocked helter-skelter north and south, east and west, as well as up and down! Their speeds of hundreds of miles per hour increased or backed up on themselves, just like a stream from a fire hose that was suddenly diverted high in the air and came crashing, splashing, down with terrific force.

Those jet streams became rivers of extremely cold air hurled far out into space to get even colder, only to return like a waterfall, plunging down through the stratosphere, and on through the troposphere; in the biosphere, they were split by the Rockies and channeled south into Hurricane Alley, where they met the warm air coming up from the Gulf, or down from the Atlantic test range. Add ten times the velocity to those currents in either direction, and you have tornadoes, the likes of which no one has ever seen. Tornadoes used to come in singles, but now they are spawned by the dozen and sometimes several combine to become one super-destructive tornado! What more proof do we need than the events of April 11, 1979, down Texas and Oklahoma way and several times since?

FIGURE 8

A low air burst showing toroidal fireball and dirt cloud. The atmospheric test bomb exploded by China on September 26, 1976, was this type. (*From* Effects of Nuclear Weapons [ENW])

The terrible part of this analogy is that we know water is heavy and cannot be compressed. The impact comes from above and in a visually impressive way. A 20 kiloton bomb was detonated well below the surface in the Bikini lagoon, which was 200 feet deep. Now this small bomb raised approximately a million tons of sea water in a column 6,000 feet across and carried a great deal of powdered coral from the bottom of the lagoon. This was deposited as a heavy dusting over all the ships in the area. Of course it was heavily radioactive!

The spray dome began to form at about four milliseconds after the detonation. It rose at roughly 2,500 feet per second, until slowed by gravity and air resistance. It rose to almost 10,000 feet. This is much less than the height of an air burst or atmospheric, radioactive cloud.

A 15,000 megaton bomb detonated in the atmosphere might just blow a livable amount of air within reach of the moon's gravitational pull. And *they* say that weapon atmospheric testing would have no more effect on the weather than a severe thunderstorm! When that air with its radioactive particulates falls back with the fountain effect of the rock in the pond, it would spread over a thousand times wider area. High winds would come straight down out of clear skies!

VAN ALLEN RADIATION BELTS

James A. Van Allen, an American physicist born in 1914, discovered the belts. While analyzing data from the rocket Explorer I, he found the lower belt. It consisted of subatomic particles, captured from the solar wind by the earth's gravity, encircling the earth about 550 miles above sea level. The outer belt, which is still believed to be intact at this time, is about 4,000 miles out. It was discovered a year later by a lunar probe. It also is composed of subatomic particles also captured from the solar winds in the same way. The inner and outer belts conform to the earth's magnetic field. Removal of either of these screens for even a few seconds exposes the earth to a terrific dose of increased solar radiation and its accompanying problems.

Scientists warned the military not to disrupt either of these protective shields. Of course, they didn't listen. It is thought that disrupting the lower Van Allen belt brought on the droughts around the world. See Fig. 6, showing an absence of clouds over North America!

A world-renown physicist believes that "cosmic rays striking the earth act as lightning rods and cause lightning to strike the earth 360,000 times an hour" (see Plate 2). Apparently the earth is a huge rotor, rotating in an electrically charged field of the atmosphere.

For the last several years, the weather has been fouled up. This time the cause has been manmade instead of due to an act of God or Mother Nature. First, the United States began testing atom bombs in the atmosphere. One hundred and ninety were admittedly sent up and detonated! This was from 1945 to 1962.

The USSR then got into the act and, of course, had to duplicate our tests and go us one better at least! This totaled about 380 punches into the troposphere and stratosphere. Heavy, heavy hangs over our heads!

Some meteorologists really believe that opening a 50- or 100-mile hole in the air ocean above us can in *no way* affect our weather! Of course, they probably didn't realize that the bombs were powerful enough to do it. But the United States and the USSR discovered that they were actually, along with the United Kingdom, loading the upper air with radioactive debris that eventually would blanket the world—and it *has!*

They agreed in 1962 to stop atmospheric testing and, apparently, they made it stick! Of course, France, China, and India were just becoming nuclear powers and had to make their own tests. Depending on where and at what altitude the bombs were detonated, every atmospheric test by anyone left its load of radioactive debris up there to thicken the blanket. So far, this is *who they are!*

WHAT *THEY* DID!

There was a nice little island in the Western Pacific, just 2° north of the equator and 157° west longitude, called Christmas Island. I say there was an island there because it was used as a testing ground by the United States and the United Kingdom for atomic weapons. I don't even know if the island ever was nice; however, most islands in the Western Pacific are nice.

The last atmospheric test the United States made was in 1962, from Johnson Island, also in the Western Pacific, 17° north and 169° west. Some time after the last test out there, it happened!

OREGON'S COLUMBUS DAY STORM

Warning No. 1

On October 12, 1962, Columbus Day, the Northwest was hit by what was called a tropical typhoon. Yet there was no warning from ships at sea! This was the first on record for the Pacific coast. No one knew it was coming because it wasn't water borne, nor did it come overland up the coast! It came *down out of the air above us!* Note the rays coming down from the upper air in Plate 3. An eye witness describes it.

> Thank God, it was a dry storm. It was not forecast by the U.S. Weather Service either. The day was hot with a light breeze blowing. It was a beautiful, typical fall day, with the trees in full green leaf. Bus loads of agricultural workers were unloading in the northwest Burnside district, on Portland's Third Avenue. The area was loaded with taverns, small groceries, and cheap hotels. I was driving a route wagon and was parked at the curb to service a small grocery. A white helmeted Civil Defense warden rushed up from down the street and said, "Say, buddy, there's a hell of a storm coming down the valley. Everybody is supposed to get off the streets."
>
> "What is this, some kind of a drill?" I asked. "How come it isn't on the radio or TV?"
>
> "Well, the only fellow that knows how to run the emergency radio system is out in the woods hunting and can't be reached. More bureaucratic planning, that's why, and hell, no, it's not a drill. It's tearing Eugene to pieces. [Eugene is Oregon's second largest city.] It must be a tornado. The TV and radio stations were definitely not prepared for an emergency of this magnitude."
>
> Suddenly, the air became deathly still. Portland was a city without a sound—*real scary.* It seemed the air was too thin to carry sound waves! Then a wonderful, sweet odor like in the tropics, or a greenhouse, filled the air. [What other proof do we need of its origin?]
>
> "Warden," I asked, "are you wearing perfume?"
>
> "Nope, not me. It's the air, I guess. Sure smells sweet, doesn't it?"

"Man, oh, Man, that's the sweetest air I ever smelled," I said. We both looked up at an empty six-story building across the street at the sound of breaking glass. A top story window had actually blown out. A rain of glass landed on the sidewalk. The glass pane was completely gone from the sash! Not a shard of glass remained in the frame. I went over to look at it. There, on the sidewalk, lay the window pane, in tiny pieces like broken safety glass. We thought at first that it was just a wino pitching a bottle through the window. However, there was no bottle around, and it should have only made a hole in the window. That spurred the warden to try to herd the winos and others off the streets. The winos were determined to get their jug, as they had been doing hard farm work all day. Others scurried for shelter. I returned to my truck just as a bus load of workers skidded to a stop in the middle of the intersection.

The doors flew open and the driver lit, running. Workers poured out and one ran white-faced over to the wagon man, an acquaintance of his, always good for something to eat when cash was short, and said, "Hey, buddy, do you know what this is." He held out a piece of thick glass as big as his hand. "This is a piece of the windshield out of that bus. That's why the driver stopped, opened the doors and lit running! A bolt of lightning came in through the windshield, down the aisle and out the back. See, it's all melted on the edges." It was!

We looked around toward the sound of what we thought was a collision. Down the street a block, parked at the curb, sat a new Ford convertible flat on its frame—squashed—heaped high with bricks from the rooftop chimney of the six-story American Hotel [see Fig. 9].

High winds were rapidly coming down to the street level. The air pressure had dropped so low that windows everywhere exploded out. This opened up buildings and when seconds later the wind hit the street level, it took off roofs in many places all over the city. Some it stacked across the street, some were carried blocks away. Where there were holes or tunnels underground, first they blew out then they caved in. The air was literally coming out of the ground!

Huge sheets of glass and corrugated roofing were actually floating several feet off the ground in the air, as though on a stream of water and going north with the wind. It got dark rapidly. All power was off except for feeder lines lying in the street, writhing around like a snake spitting fire and sparks. Signboard uprights made to withstand any wind were twisted around like corkscrews. Trucks and trailers were blown up against bridge railings. The wind had died down to a light breeze when I decided to head for home.

The Burnside Bridge, a six-lane cement and steel Bascule type, had one leaf raised a foot by the wind pressure, this above all resistance from gears, etc. Trying to get home after the wind died down, I headed the truck across the bridge. A warden ran out to flag me down and threw his hands up in despair as I kept on going. I didn't consider for a minute that any windstorm could damage a bridge that solid. There was not a light anywhere, only the headlights of the truck high in the air going up the bridge ramp. As I came to the center of the bridge, I saw in a split second there was no sign of the opposite bridge leaf. I had just slid back the truck door and was expecting to have to swim for it when—WHAM—the front wheels dropped down and hit the other leaf like a ton of bricks. The east side leaf was a full foot lower than the one I was on. It darn near busted my dentures that time! Later it was announced that the bridge wind gauge registered 150 mph. before it went off the scale. It was only a light breeze as I crossed the bridge. If I had been caught on the bridge in my light truck, the storm would have blown me—truck and all—off for sure. With all lights out, how would a person find their way to the river bank? Man! Oh, man, it was absolutely pitch dark, like a block of unmelted asphalt! No stars, no moon, no glow of city lights, it was absolutely *black*! A first time for me when *no* light of *any kind* was visible *anywhere!* WOW! Was that ever a storm! I've been in typhoons before in the Philippines, but never one like that. The fact that it only lasted about four hours seems to prove it wasn't ocean borne.

FIGURE 9

A new sports car demolished by falling bricks from a six-story building in Portland, Oregon, during the Columbus Day Storm. (*From* The Big Blow *by Ellis Lucia; used with permission.)*

Warning No. 2

This storm was a warning of things to come! Perhaps one or more of those 3,000-foot tropospheric waves had crashed down on the east side of the coast range. It landed in the north of California then it leapfrogged the Siskiyou Mountains and landed squarely on the city of Eugene, at the upper end of the Willamette Valley. It then was funneled down the valley as a huge river of air! However, instead of flowing quietly as a river does, it came down like a dam bursting! If warnings could have been given, and people had known enough to open buildings to relieve the internal pressure, there would have been far less damage. Houses with fireplaces, with the damper open did not

FIGURE 10

The 1,000-year-old Clatsop Fir, considered the world's largest, weakened by the Columbus Day Storm, fell a week later in a less-severe storm. (*From* The Big Blow *by Ellis Lucia; used with permission*)

lose a single window. This has also proved to work during tornadoes! It was *not a wave!* Evidently, driven by jet streams, kicked around in the tropics by atmospheric testing, a lot of sweet biospheric tropical air was sucked up by the vacuum in the mushroom stem of an atmospheric fusion bomb! Oregon's Columbus Day Storm had to come from the Johnson Island bomb testing! Trees, big ones, were felled by the thousands (see Fig. 10).

Utility crews were called in from neighboring states. Washington was also hit, but Oregon's Willamette Valley took the brunt of the blow, which followed the river like a huge funnel, moving counterclockwise as a good northern vortex should. Fig. 11 shows the path of the storm.

Train loads of glass were sent express from many factories and warehouses all over the nation to Oregon and Washington. There were millions of dollars in damages. Some insurance companies went bankrupt. Sixty days later things were still not back to normal. Insurance companies are still struggling to recover from their losses. The summary figures below, taken from *The Big Blow: The Story of the Pacific Northwest's Columbus Day Storm,* by Ellis Lucia, give an idea of what happened that day. (The Columbus Day Storm was declared the nation's worst natural disaster of 1962 by the Metropolitan Life Insurance Company.)

Date of the Big Blow—Oct. 12, 1962

Size of storm path area in N. California, Oregon, Washington—corridor 125 mi. wide, 1,000 mi. long, 75,000 sq. miles.

Barometric low point 28.41

Peak wind velocities:

Hebo	170 mph
Corvallis	125 mph
Portland	116 mph

Length of storm passage over any point—About 2 hours

Total dead in 3 states	48
Oregon Fatalities	24
Damage:	
Oregon	$170,000,000
Washington	$ 40,000,000
Total	$210,000,000

Multnomah	$169,800.00
Washington	$393,178.00
School damage, Washington State	$56,740.52

Damages, losses to cities:
(public facilities)

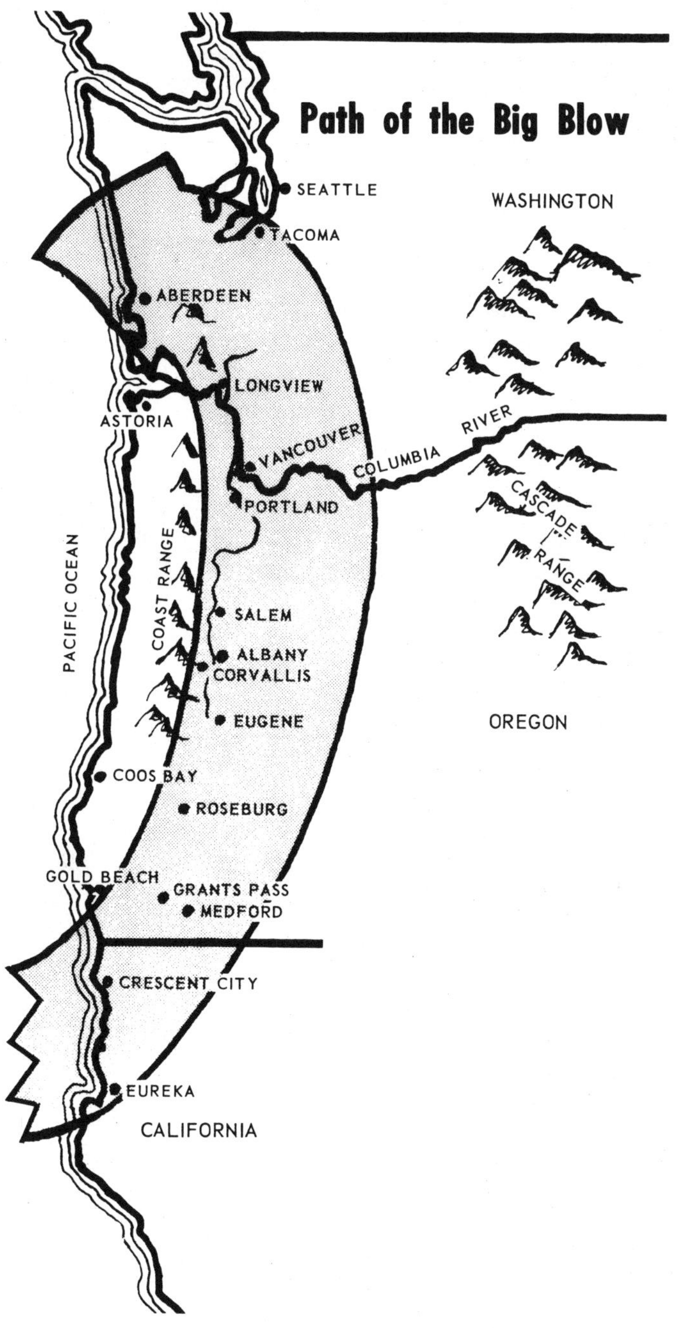

FIGURE 11

The crescent-shaped, counter-clockwise motion of the Columbus Day Storm path. (*From* The Big Blow *by Ellis Lucia; used with permission*)

No. families without power	496,000
No. telephones out	129,900
Insurance claims:	
Oregon	70,000
Amnt. paid	$20,000,000
Washington	28,216
Paid (87% of total or 24,676 claims)	$4,230,200
Airport destruction in Oregon:	
(13 airports)	
Planes damaged	226
Totally wrecked	56
Hangars wrecked	24
Est. total damage	$1,649,130
School damage (21 Oregon counties):	
Total	$2,250,418
Heaviest hit schools:	
Clackamas County	$150,000.00
Curry	$240,000.00
Lane	$289,000.00
Lincoln	$275,000.00
Linn	$229,000.00
Marion	$254,000.00
Portland—	
Total (est.)	$1,700,000
Bur. of Parks	$875,000
Dock comm.	$225,000
Dept. Publ. Wks.	$525,000
Other depts.	$55,709
Est. trees down	16,000
Park tree destruction	4,000
No. stump removals	2,794
Salem	$220,000
Corvallis	$20,000
Eugene	$50,000
Oregon City	$35,000
Albany	$27,000
Hillsboro	$30,000
Cost of removal	$200,000
Longview, Wash.	$123,764
Vancouver, Wash.	$100,000
Farm Damage:	
(17 Oregon Counties)	$61,300,000
Forest Blow-down:	
Oregon	6,282,391,000 b.f.
Washington	4,914,425,000 b.f.
Total	11,196,816,000 b.f.
Loss estim. after salvage	1,500,000,000 b.f.

Many utility poles and lines went down with the trees. Deciduous trees that weren't blown over or broken off, in many places, especially open parks, looked as if a giant hand had grasped the leafy part and wrung them dry, twisting them like a dishrag. The formerly bright green leaves were now dark brown —cooked—all moisture sucked out. New Yorkers and Long Islanders may remember that after their so-called hurricane, beautiful oaks and other trees along the coast looked the same way. *They* said it was windblown salt water, remember? How many years have those trees had salt water spray blown on them and *never* looked like that? What's their excuse in the Willamette Valley? The East Coast storm came down out of the air the same as the Columbus Day Storm, but from the Atlantic test range! Now back to Oregon. It's still with us. It may never stop!

VIOLENT WIND, HAIL SLAM SANDY AREA was the headline of a story by Nancy Barker, *Oregon Journal* correspondent:

> Sandy Oregon—Citizens here Tuesday, August 21, 1979, began cleaning up the aftermath of a sudden, violent thunderstorm that brought wind of up to ninety miles an hour and hail stones the size of golf balls Monday evening, causing severe damage to several homes but no reported injuries.
>
> City streets "look like they are covered with green snow," said Sandy police officer Rick Peterson Tuesday. He said tree limbs and debris are everywhere.
>
> Peterson said the storm was "peculiar, with the wind going in a circular direction." He said that limbs were broken off in one direction on one side of a street and in the other direction on the other side.
>
> Doug Gaylord, a golf course employee at the Mountain View Golf Course in nearby Boring said ten members out on the course came running to the clubhouse. "The storm really scared us," they said.

THE YEAR OF THE TORNADOES

On April 3, 1974, meteorologists of the U.S. Weather Service noted a low-pressure system of cold air moving eastward from the Pacific coast as a gigantic vortex covering several states. It was rotating counterclockwise, as it should. Satellite pictures also showed that a mass of warm air was moving up from the Gulf of Mexico and rotating clockwise! This was as it should be if it occurred below the equator. However, it is a no-no in the Northern Hemisphere. It's against the law of nature. Could this clash of opposing vortexes have been caused by the French atom bomb tests at their desert testing area in Africa? Hardly. Our own tests of the Argus Operation in the South Atlantic were responsible! This clash of hot and cold air, each spinning a different way, sent tornadoes too numerous to count raging across the country from Huntsville, Alabama, to Michigan.

Winds of over 300 mph blew the anemometer off the scale! Hundreds of people were killed and thousands were injured. Damage in the millions also occurred. The equatorial jet stream was evidently blasted north of the equator with enough force to carry it clear up into Michigan (see Fig. 12). These opposing vortexes triggered the tornadoes. What more proof do we need than the hell raised in Hurricane Alley in April 11, 1979, down Texas and Oklahoma way, or in Connecticut, October 4, 1979, and numerous other places since? This was the second warning.

FIGURE 12

Area hit by tornadoes in the spring of 1974. (*From* Reader's Digest, *with permission of George Buctel*)

GOING UNDERGROUND—A SENSIBLE SOLUTION

Here I will digress a bit, though it is too late to help those people who have already suffered devastation. I suggest that people in the Hurricane Alleys of the world, wherever they may be, scrape the surface of their property clean of all debris. With heavy equipment, excavate the front as deep as is required. Then build all homes and businesses *underground* and retain parks, gardens, and farms on the surface! Keep the streets and sidewalks as they are. Montreal, Canada, has just such an underground city with rubber-tired mass transit. Think of the energy saved and the lives that will be saved in the future with no need for

PLATE 1

The Medal of Honor, awarded in the name of Congress for conspicuous gallantry and valor at the risk of life above and beyond the call of duty. *(Courtesy U.S. Navy, Photo Section)*

PLATE 2

Lightning over Tampa Bay. A world-renown physicist believes that cosmic rays striking the earth act as lightning rods. *(Photo by Nelson Medina, Photo Researchers, Inc., New York, New York)*

PLATE 3

A tropical typhoon hit Oregon on October 12, 1962, with no warning because it was neither water nor land borne; it descended from the atmosphere! Note the rays coming down from the upper air. *(Photo by Gertrude E. Myers, courtesy Oregon Historical Society)*

Not only have they blown our weather, they are killing our oceans! Marine animals of t
and Atmospheric Administration)

hemisphere are shown above. *(Courtesy U.S. Department of Commerce, National Oceanic*

PLATE 5

On June 16, 1979, forty-one enormous sperm whales beached themselves near Florence on the central Oregon coast. The reason for this beaching has baffled scientists, although the most plausible explanation is that the whales were *cooked* to death—victims of an underwater detonation of an atomic or fusion bomb! Plates 5 to 9 are photographs of the stranded whales after dissection. *(Photos courtesy of University of Oregon Marine Science Laboratory, Charleston, Oregon)*

PLATE 6

Teeth in the whales' jaws varied from between 40 and 48 (the sperm whale is the only whale that has teeth and its teeth are in the lower jaw only). *(Photo by Larry Oglesby)*

PLATE 7

A view of a whale's upper jaw, with sockets for lower-jaw teeth. *(Photo by Larry Oglesby)*

PLATE 8
Measurement of a fetus not fully developed. The stump of the umbilicus, which was cut, can be seen. *(Photo by Margie Ryan)*

PLATE 9
Fetus with umbilicus still attached, near term. *(Photo by Michael Graybill)*

hurricane insurance—all at less cost than one disaster! Earth-covered structures will also be more or less radiation proof depending on the depth of the overlay. They will mean better living, with no need for one to leave friends or the area. The world of the year 2000, in Hurricane Alley *now!* What a tourist attraction. Remember the slogan: "If it saves energy, it will pay for itself!"

To learn more about underground architecture and to get reliable information on such structures write to Underground Space Center, University of Minnesota, 11 Mines and Metallurgy, 221 Church Street S.E., Department of Civil and Mineral Engineering, Minneapolis, Minnesota 55455.

Again, digressing a bit, if people in California's burned-out areas would go underground with their reconstruction—they already have pools and perhaps they could save a chimney, a cement slab or so—if fire came again only the brush would be above ground to burn! It would cost less to build, cool, or heat!

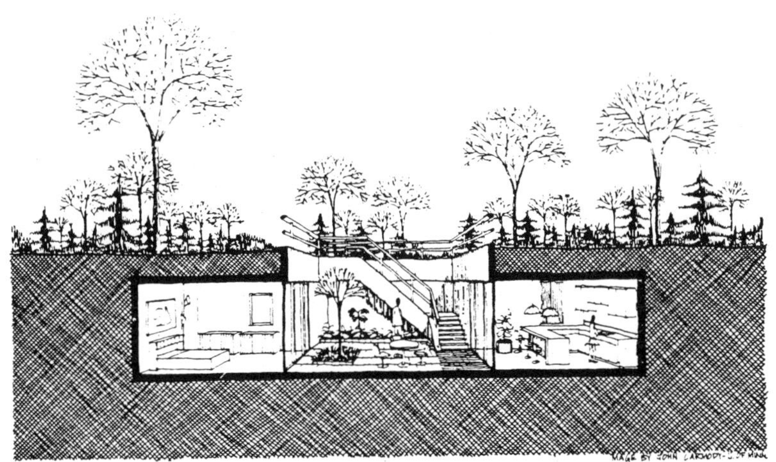

FIGURE 13

Ecology House in Marstons Mills, Massachusetts, designed by architect John E. Baynard, Jr., AIA. *(Used with permission)*

FIGURE 14

FIGURE 15

Exterior and interior views of Ecology House, by John E. Baynard, Jr., AIA. *(Courtesy Underground Space Center, University of Minnesota)*

Just think! No fire insurance, plus radiation fallout protection! Then reseed or plant the hills.

RADIOACTIVE PARTICULATES

While you're at it, if it's possible, dig or drill a well of your own. The HEW (Health, Education and Welfare) seems determined to cram fluorides into our bones and endocrine glands by adding it to the drinking water, along with chlorine and radioactive fallout. What do you think produces hailstones the size of golf balls or baseballs? Our city reservoirs, open to the clear blue sky, must be covered! What produced the rubber-legged cows of Trout Dale, Oregon? Why are cows taken off of forage feed during periods of heavy fallout? *Cover your reservoirs!*

Environmental News, an EPA (Environmental Protection Agency) publication, reported Friday, March 24, 1978:

> Fallout has been reported in rainwater samples by Colorado, Arkansas, and Missouri. Measurable radiation fallout was indicated in Wyoming, Texas, Oklahoma, Arkansas, Missouri, Mississippi, Tennessee, Wyoming, with highest levels in Wyoming and Colorado. Colorado State Department of Health recommended that persons refrain from drinking rainwater and melted snow for the next several days. This also applied to dairy cows for a week!

The interaction of storm fronts with radioactive, debris-laden clouds acts as cloud seeding! This in Oregon came down as rain. Other states got it as snow. Some had softball-size hailstones. The tiny nuclides gather ice on the way down, and the descent starts thousands of feet higher than normal clouds. No wonder the hailstones arrive jumbo size! During the next hailstorm, check the stones with a Geiger counter or dosimeter. It will probably not be sensitive enough, but give it a try anyhow. Oregon got it mild!

Tritium

The interaction of fast neutrons in cosmic rays with nitrogen nuclei in the air leads to the formation of tritium in the normal atmosphere; this radioactive isotope of hydrogen has a half life of about 12.3 years. Small amounts of tritium are formed in fission but larger quantities result from the explosion of thermonuclear weapons. The fusion of deuterium and tritium proceeds much more rapidly than the other thermonuclear reactions.

Most of the tritium remaining after a nuclear explosion, as well as that produced by cosmic rays, is rapidly converted into tritiated water, HTO; this is chemically similar to ordinary water (H_2O) and differs from it only in the respect that an atom of the radioactive isotope tritium (T) replaces one atom of ordinary hydrogen (H). If the tritiated water should become associated with natural water, it will move with the latter.

The amount of tritium on earth, mostly in the form of tritiated water, attained a maximum in 1963, after atmospheric testing by the United States and the USSR had ceased. The tritiated water in the troposphere is removed by precipitation and at times, in 1958 and 1963, following extensive nuclear weapons test series, the tritiated water in rainfall briefly reached values about 100 times the natural concentration.

In these reports the first reading on tritium was taken on December 5, 1975. It seems that the detectable level listed at 450 pCi/liter was seldom reached. In June of 1977 it jumped to 530 pCi/liter at La Grande; of course, this was a thirty-day reading. At Astoria, west of the Cascade Range, no readings were taken from November 17, 1972, to March of 1973. From March to June 27, 1973, the tritium reading jumped to 1,000! Other hot spots, or accumulations, could have been overlooked.

Tritium, carbon-14, and cesium-137 are in a different category. They are distributed throughout the body and so can cause irradiation of the gonads. Soft, low-energy nuclides *are* dangerous!

Gross Beta

Carbon-14, with a half life of 5,730 years emits beta particles of low energy but no gamma rays! The RBE (relative biological effectiveness) depends on whether it's internal or external! According to the study prepared by the U.S. Departments of Defense and Energy:

> All radiations capable of producing ionization (or excitation), directly or indirectly, e.g., alpha and beta particles, tritum,

> X-rays, gamma rays, and neutrons, cause injury of the same general type. Although the effects are qualitatively similar, the various radiations differ in the depth to which they penetrate the body and in the degree of injury corresponding to a specified amount of energy absorption.* [See Figs. 34, 35, 36, 37, 38, 39, 40, 41 for external beta burns from early fallout.]

Internal: Beta emitting particles, leaving the lungs by way of the lymphatic system, tend to accumulate principally in the tracheobronchial lymph nodes, thereby leading to an intense localized radiation dose. *Chronic coughing, voice change, intense salivation followed by malaise, either preceded or followed by vomiting,* can carry on for a month or so depending on the individual. Doctors seem able to prescribe for the coughing, which may give two or three weeks of relief, but a return bout usually hits and if not overcome may take a month or two and leave the victim dangerously weak. It is frequently diagnosed as flu!

However the special interest in the delayed fallout arises from the fact that it may occur in significant amounts in many parts of the globe remote from the point of the nuclear detonation. Regardless of its mode of entry—inhalation, ingestion, or wounds—cesium-137 is soon distributed fairly uniformly throughout the body as a whole. Because of this, the entire body would be irradiated by both beta particles and gamma rays.

After a detonation, beta readings bounce up and down a bit as the debris clouds make repeated passes around the globe. Thunderstorms scrounge radioactive nuclides from the stratosphere and the upper troposphere, resulting in rain and snow-outs!

Gross Alpha

This is the activity of the fissionable material, part of which remains after an atomic explosion. The fissionable uranium and plutonium isotopes are radioactive. Their activity consists of the emissions of what are called "alpha particles." These are a form of nuclear radiation.

Because of their greater mass and positive electrical charge,

**The Effects of Nuclear Weapons,* eds. Samuel Glasstone and Philip J. Dolan (Washington D.C.: U.S. Printing Office, 1977), p. 576.

alpha particles suffer early fallout from the radioactive debris clouds. Alpha particles are much less penetrating than beta particles or gamma rays of the same energy.

It is doubtful that these alpha particles can get through unbroken skin. Of course, if your skin itches and you scratch you may break the skin. They cannot penetrate clothing! The uranium or plutonium present in weapon residues (fallout) is not a hazard outside the body. However, if it enters the body through skin abrasions, inhalation, or ingestions of drinking water, which is what we are talking about here, the effects may be disastrous! Of course, humanity may be more resistant to radiation than laboratory rats. We've come a long way, but then so have the rats!

Alpha emitters, e.g., plutonium, beta particles of soft (low) energy, and gamma ray emitters can deposit their entire energy within a small, possibly single, sensitive volume of body tissue, causing considerable damage. There seems to be a rash of liver cancer across the country!

Below are reports of tests made on rainwater in an Oregon statewide sampling program to determine the amount and concentration of radionuclides and their radioactivity. Laboratories around the world were alerted as soon as the first atmospheric bomb was detonated by the People's Republic of China. Three types of radioactive particulates were monitored consistently: gross alpha, gross beta, and tritium. The data here were furnished by George L. Toombs, Supervisor, Environmental Radiation Surveillance Program, Radiation Control Section for the state of Oregon. Mr. Toombs states:

> In comparing the data, the differences in sampling frequency should be taken into consideration. The Portland location sampling consisted of a two-week composite sample, except during periods of high fallout when more frequent samplings were performed to follow the passage of the intrusion. At Klamath Falls and La Grande, the sampling was on a monthly basis. The remaining stations are on a quarterly schedule.
>
> Since fallout contains both short-lived and long-lived activity, the Portland data collected more frequently is the most useful when defining intrusions of short-lived activity. The other locations on a more extended sampling interval denote primarily the longer-lived activity.

TABLE 3

Radioactivity and Concentrations of Radionuclides in Rainwater

Location: Portland (State Office Building)

	Concentrations expressed as pCi/liter		
Collection Period	*Gross Alpha*	*Gross Beta*	*Tritium*
January 8-14, 1971	<1	19	
January 15-21, 1971	<1	79	
January 22-28, 1971	—	23	
January 29-February 4, 1971	2	454	
February 5-11, 1971	1	76	
February 12-18, 1971	<1	76	
February 19-25, 1971	<1	30	
February 26-March 4, 1971	<1	68	
March 5-11, 1971	<1	191	
March 12-18, 1971	<1	89	
March 19-20, 1971	<1	117	
March 26-31, 1971	<1	135	
April 2-8, 1971	1	198	
April 8-15, 1971	<1	60	
April 16-22, 1971	<1	167	
April 23-29, 1971	1	299	
April 30-May 6, 1971	2	502	
May 7-13, 1971	1	210	
May 14-20, 1971	2	327	
May 21-27, 1971	4	173	
May 28-June 3, 1971	2	275	
June 4-10, 1971	2	327	
June 11-17, 1971	<1	111	
June 18-24, 1971	1	240	
June 25-July 1, 1971	<1	74	
July 2-9, 1971	<1	64	
July 10-16, 1971	<1	76	
August 16-20, 1971	1	73	
August 21-27, 1971	<1	59	
August 28-September 3, 1971	<1	58	
September 4-10, 1971	<1	64	

TABLE 3—*Continued*

Collection Period	*Gross Alpha*	*Gross Beta*	*Tritium*
	Concentrations expressed as pCi/liter		
September 11-17, 1971	<1	—	
September 25-October 1, 1971	2	53	
October 9-15, 1971	9	39	
October 16-22, 1971	<1	33	
October 23-29, 1971	<1	37	
October 30-November 5, 1971	<1	—	
November 6-12, 1971	<1	—	
November 13-19, 1971	<1	32	
November 20-26, 1971	<1	80	
November 27-December 3, 1971	<1	133	
December 4-10, 1971	1	53	
December 11-17, 1971	<1	76	
December 18-24, 1971	1	17	
December 24-31, 1971	<1	43	
January 8-14, 1972	<1	14	
January 15-21, 1972	<1	252	
January 22-28, 1972	2	250	
February 5-11, 1972	<1	74	
February 12-18, 1972	<1	52	
February 19-25, 1972	<1	42	
February 26-March 3, 1972	<1	37	
March 4-10, 1972	<1	43	
March 11-17, 1972	<1	10	
March 20, 1972	—	55	
March 22, 1972	5	854	
March 23, 1972	61	9146	
March 24, 1972	25	4484	
March 27, 1972	6	1691	
March 28, 1972	2	1731	
April 3, 1972	3	511	
April 4, 1972	3	—	
April 5, 1972	9	—	
April 6, 1972	1	224	
April 7, 1972	<1	62	
April 10, 1972	<1	61	

TABLE 3—*Continued*

Collection Period	*Gross Alpha*	*Gross Beta*	*Tritium*
	Concentrations expressed as pCi/liter		
April 11-14, 1972	<1	98	
April 15-21, 1972	2	91	
April 22-27, 1972	<1	116	
April 28-May 5, 1972	<1	144	
May 6-12, 1972	<1	17	
May 13-19, 1972	<1	285	
May 20-26, 1972	<1	174	
June 10-16, 1972	<1	207	
June 24-30, 1972	<1	49	
July 8-14, 1972	<1	—	
August 12-18, 1972	<1	41	
August 19-25, 1972	<1	91	
September 18-22, 1972	<1	10	
October 7-13, 1972	1	—	
October 21-27, 1972	<1	7	
October 28-November 3, 1972	<1	8	
November 4-10, 1972	<1	11	
November 11-17, 1972	<1	72	
November 18-24, 1972	<1	—	
November 25-December 1, 1972	<1	29	
December 9-15, 1972	<1	4	
December 16-22, 1972	<1	4	
December 30-January 5, 1972	<1	6	
January 6-12, 1973	<1	6	
January 13-19, 1973	<1	11	
January 20-27, 1973	<1	6	
January 27-February 2, 1973	<1	33	
February 10-16, 1973	<1	13	
February 17-23, 1973	<1	14	
February 24-March 2, 1973	<1	6	
March 3-9, 1973	<1	3	
March 10-16, 1973	<1	11	
March 17-23, 1973	<1	9	
March 24-30, 1973	1	15	
March 31-April 6, 1973	<1	8	

TABLE 3—*Continued*

	Concentrations expressed as pCi/liter		
Collection Period	*Gross Alpha*	*Gross Beta*	*Tritium*
April 14-20, 1973	<1	8	
May 1-31, 1973	<1	11	
June 12-18, 1973	1	8	
June 23-26, 1973	<1	—	
August 1-3, 1973	2	66	
August 24-25, 1973	<1	3	
September 7, 1973	1	6	
September 18-24, 1973	<1	9	
September 25, 1973	<1	5	
October 15-22, 1973	8	2	
October 23-27, 1973	<1	5	
October 30-November 5, 1973	<1	5	
November 14-21, 1973	<1	3	
November 22-30, 1973	<1	7	
December 1-10, 1973	<1	13	
December 11-17, 1973	<1	13	
December 18-24, 1973	<1	6	
December 25-31, 1973	<1	9	
January 11-18, 1974	<1	11	
January 19-28, 1974	<1	20	
January 29-February 5, 1974	<1	13	
February 21-28, 1974	<2	18	
March 1-11, 1974	<1	36	
April 5-12, 1974	<1	73	
April 22-26, 1974	<1	64	
April 27-May 10, 1974	<1	36	
May 10-17, 1974	<1	36	
May 18-24, 1974	<1	30	
May 25-31, 1974	2	21	
June 1-7, 1974	<1	55	
June 26-27, 1974	<1	25	
July 4-5, 1974	<1	44	
July 13-19, 1974	<1	7	
August 17-23, 1974	1	11	
September 7-13, 1974	<1	13	
September 28-October 4, 1974	1	9	

TABLE 3—*Continued*

Collection Period	*Gross Alpha*	*Gross Beta*	*Tritium*
	Concentrations expressed as pCi/liter		
October 5-11, 1974	2	9	
October 19-25, 1974	<1	14	
October 31, 1974	1	10	
November 2-8, 1974	<1	22	
November 9-15, 1974	<1	10	
November 16-22, 1974	1	11	
November 23-29, 1974	<1	7	
November 30-December 6, 1974	1	20	
December 6-13, 1974	2	13	
December 13-20, 1974	3	9	
December 20-27, 1974	2	13	
December 27, 1974-January 3, 1975	1	21	
January 3-10, 1975	<1	12	
January 10-17, 1975	<1	17	
January 17-24, 1975	1	22	
January 24-31, 1975	3	6	
January 31-February 7, 1975	<1	22	<450
February 7-14, 1975	<1	29	
February 14-21, 1975	<1	16	
February 21-28, 1975	<1	21	
February 28-March 7, 1975	3	—	<450
March 7-14, 1975	<1	32	
March 14-21, 1975	2	20	
March 21-28, 1975	3	15	
March 28-April 4, 1975	<1	72	<450
April 4-11, 1975	<1	52	
April 11-18, 1975	1	38	
April 18-25, 1975	2	13	
April 25-May 2, 1975	<1	41	<450
May 2-8, 1975	<1	6	
May 9-16, 1975	<1	12	
May 16-23, 1975	<1	30	
May 23-30, 1975	2	55	<450
June 13-20, 1975	—	85	
June 20-26, 1975	<1	13	

TABLE 3—*Continued*

	Concentrations expressed as pCi/liter		
Collection Period	*Gross Alpha*	*Gross Beta*	*Tritium*
June 26-July 3, 1975	<1	19	<450
July 26-August 1, 1975	3	11	
August 2-8, 1975	2	13	
August 15-22, 1975	1	<1	
August 22-29, 1975	1	3	
August 29-September 2, 1975	1	12	
October 1, 1975	1	2	
October 10-16, 1975	12	4	
October 17-24, 1975	4	4	
October 24-29, 1975	<1	2	
October 30-November 6, 1975	1	3	
November 7-14, 1975	<1	1	
November 14-21, 1975	3	6	
November 21-28, 1975	<1	1	
November 28-December 5, 1975	1	2	<450
December 5-12, 1975	<2	3	
December 12-19, 1975	<2	4	
December 19-26, 1975	<2	4	
January 2-9, 1976	<2	1	<450
January 9-16, 1976	<2	1	
January 16-23, 1976	2	6	
January 23-30, 1976	3	2	
January 31-February 6, 1976	<2	2	
February 6-13, 1976	<2	3	
February 13-20, 1976	<2	2	
February 20-27, 1976	<2	1	
March 5-12, 1976	<2	7	<450
March 12-19, 1976	<2	3	
March 19-26, 1976	<2	2	
March 26-April 2, 1976	—	4	
April 2-9, 1976	<2	6	
April 9-16, 1976	3	6	
April 16-23, 1976	<1	4	
April 30-May 7, 1976	<2	7	
May 7-14, 1976	<2	2	
May 14-28, 1976	<1	12	<450

TABLE 3—*Continued*

Collection Period	*Gross Alpha*	*Gross Beta*	*Tritium*
	Concentrations expressed as pCi/liter		
May 28-June 4, 1976	<2	2	
June 4-11, 1976	—	12	<450
June 12-19, 1976	1	4	
June 25-July 2, 1976	—	1	
July 2-9, 1976	<2	3	<450
July 9-16, 1976	<2	2	
July 30-August 6, 1976	<3	7	
August 6-13, 1976	<2	1	<450
August 13-20, 1976	<1	1	
August 20-27, 1976	<1	2	
September 3-10, 1976	—	2	
September 10-17, 1976	—	9	
September 17-24, 1976	—	4	<450
September 17-October 2, 1976	—	329	
October 2-10, 1976	14	128	
October 10-11, 1976	3	16	
October 11-24, 1976	2	468	
October 24-25, 1976	4	129	
October 25-26, 1976	6	233	
October 26-29, 1976	1	162	<450
October 29-November 1, 1976	1	80	
November 1-2, 1976	—	264	
November 2-9, 1976	—	292	
November 9-16, 1976	<1	41	
November 16-18, 1976	—	33	
November 18-22, 1976	—	69	<450
November 22-25, 1976	—	68	
November 25-December 7, 1976	—	237	<450
December 7-10, 1976	2	18	
December 10-21, 1976	—	48	
December 21-23, 1976	<1	16	
December 23-30, 1976	<1	7	
December 30, 1976-January 7, 1977	<1	12	
January 7-14, 1977	<1	14	
January 14-21, 1977	<1	11	

TABLE 3—*Continued*

	Concentrations expressed as pCi/liter		
Collection Period	*Gross Alpha*	*Gross Beta*	*Tritium*
January 21-February 4, 1977	<1	494	
February 4-11, 1977	<1	32	
February 11-18, 1977	<1	15	
February 18-25, 1977	<1	4	
February 25-March 4, 1977	<1	3	
March 4-11, 1977	<1	9	
March 11-18, 1977	<1	4	
March 18-25, 1977	<1	9	
March 25-April 1, 1977	1	15	
April 1-8, 1977	2	21	
April 8-15, 1977	1	36	
April 15-22, 1977	2	41	
April 22-29, 1977	5	26	
April 29-May 6, 1977	2	29	
May 6-13, 1977	14	25	
May 13-20, 1977	1	15	
May 20-27, 1977	<1	29	
May 27-June 3, 1977	<1	34	
June 3-10, 1977	6	91	
July 1-10, 1977	<1	21	
July 15-22, 1977	1	25	
July 22-29, 1977	<1	30	
August 19-26, 1977	—	49	
September 2-9, 1977	—	12	
September 9-16, 1977	—	158	
September 16-19, 1977	—	11	
September 19-20, 1977	—	21	
September 20-21, 1977	—	12	
September 21-22, 1977	—	57	
September 22-23, 1977	—	5961	
September 23-24, 1977	—	1587	
September 24-25, 1977	—	469	
September 25-26, 1977	—	278	
September 26-28, 1977	—	412	
September 28-29, 1977	—	370	
September 29-30, 1977	—	425	
September 30-October 3, 1977	—	100	

TABLE 3—*Continued*

Collection Period	*Gross Alpha*	*Gross Beta*	*Tritium*
	Concentrations expressed as pCi/liter		
October 3-7, 1977	—	265	
October 7-10, 1977	—	56	
October 10-13, 1977	—	895	
October 28-November 4, 1977	—	16	
November 4-12, 1977	—	15	
November 12-18, 1977	—	22	
November 18-25, 1977	—	21	
November 25-December 2, 1977	<1	6.3	
December 2-9, 1977	<1	14	
December 9-16, 1977	<1	8	
December 16-23, 1977	<1	23	
December 23-30, 1977	<1	21	
January 6-13, 1978	<1	12	
January 13-19, 1978	1	6	
January 19-20, 1978	<1	28	
January 20-25, 1978	<1	76	
January 25-26, 1978	4	72	
January 26-30, 1978	4	47	
January 30-31, 1978	<1	53	<450
January 31-February 3, 1978	<1	10	
February 3-7, 1978	<1	10	
February 7-10, 1978	<1	19	
February 10-17, 1978	<1	13	
February 17-24, 1978	<1	72	
March 3-10, 1978	<1	11	
March 10-14, 1978	<1	14	
March 14-23, 1978	<1	54	
March 23-24, 1978	—	70	
March 24-27, 1978	<1	19	
March 27-April 4, 1978	<1	52	
April 4-14, 1978	<1	23	
April 14-18, 1978	<1	19	
April 18-25, 1978	<1	36	
April 25-May 2, 1978	2	31	
May 2-12, 1978	2	39	
May 12-19, 1978	<1	9	

TABLE 3—*Continued*

Collection Period	*Gross Alpha*	*Gross Beta*	*Tritium*
	Concentrations expressed as pCi/liter		
May 19-28, 1978	<1	31	
May 28-31, 1978	<1	19	
June 2-9, 1978	<1	13	
June 9-16, 1978	<1	12	
June 16-23, 1978	<1	49	
July 1-7, 1978	<1	32	
July 15-16, 1978	<1	13	
August 11-18, 1978	<2	9	
August 18-25, 1978	<2	11	
September 1-8, 1978	<2	6	
September 8-15, 1978	<2	10	
September 15-22, 1978	<1	27	
September 22-30, 1978	—	31	
October 20-24, 1978	<1	24	
October 28-29, 1978	2	7	
October 29-November 3, 1978	<1	5	
November 3-10, 1978	<1	6	
November 10-17, 1978	<1	5	
November 17-21, 1978	2	2	
November 24-December 1, 1978	<1	5	
December 1-8, 1978	<1	3	
December 8-12, 1978	<1	6	
December 15-18, 1978	—	22	
December 18-20, 1978	<1	15	
December 20-21, 1978	2	11	
December 21, 1978	<1	86	
December 21-23, 1978	<1	74	
December 23-24, 1978	<1	16	
December 24, 1978-January 3, 1979	—	49	
January 3-5, 1979	—	122	
January 5-8, 1979	—	22	
January 8-12, 1979	<1	4	
January 12-16, 1979	<1	10	
January 16-19, 1979	<1	6	
January 19-23, 1979	<2	25	

TABLE 3—*Continued*

Collection Period	*Gross Alpha*	*Gross Beta*	*Tritium*
	Concentrations expressed as pCi/liter		
January 23-February 9, 1979	<2	3	
February 9-16, 1979	<1	5	
February 16-23, 1979	<1	2	
February 23-March 2, 1979	<1	6	
March 2-9, 1979	<1	4	
March 9-16, 1979	2	6	
March 16-23, 1979	<1	5	
March 23-April 6, 1979	<1	14	
April 6-13, 1979	<1	7	
April 13-20, 1979	1	7	
April 20-27, 1979	<1	13	
April 27-May 4, 1979	<1	3	
May 4-11, 1979	<1	3	
May 11-29, 1979	<1	10	
May 29-June 8, 1979	<1	10	
June 8-22, 1979	<1	3	
June 22-July 6, 1979	<1	2	
July 6-10, 1979	<1	9	
July 10-August 17, 1979	1	8	<450
August 17-21, 1979	<1	4	
August 21-31, 1979	<1	13	
August 31-September 7, 1979	<1	2	
September 7-8, 1979	<1	2	
September 8-28, 1979	<3	21	
September 28-October 19, 1979	<1	2	
October 19-26, 1979	<1	2	
October 26-November 2, 1979	<1	2	
November 2-9, 1979	<1	2	
November 9-16, 1979	2	10	
November 16-23, 1979	<1	2	
November 23-27, 1979	<1	2	
November 27-30, 1979	<1	8	
November 30-December 4, 1979	<1	1	
December 4-11, 1979	<1	3	
December 11-21, 1979	<1	2	

TABLE 4

Radioactivity and Concentrations of Radionuclides in Rainwater

Location: Bull Run Headworks* (about 35 miles east of Portland)

	Concentrations expressed as pCi/liter		
Collection Period	*Gross Alpha*	*Gross Beta*	*Tritium*
December 18, 1972-June 27, 1973	<1	5	—
June 27-December 19, 1973	<1	2	<450
December 19, 1973-March 18, 1974	<1	6	<450
March 18-August 8, 1974	2	11	<450
August 8-December 19, 1974	<1	4	<450
December 19, 1974-March 27, 1975	<1	6	<450
March 27-June 25, 1975	<1	10	<450
June 25-December 1, 1975	<1	2	<450
December 1, 1975-March 31, 1976	—	—	<450
March 31-August 12, 1976	<1	1	<450
August 12-November 9, 1976	<1	13	<450
November 9-24, 1976	—	30	<450
November 24, 1976-June 9, 1977	—	7	<450
June 9-October 27, 1977	1	15	—

*This is a very comforting report. Evidently when the debris clouds rose to cross the Cascades, they did not drop much fallout on the vast watershed of Mount Hood on the western side, the area that supplies the metropolitan area with its drinking water. Most of whatever came down must have stuck to the timber. The gross alpha went from 1 to 2, only twice. However the Bull Run reservoir did register ups and downs from December 18, 1972, to November 16, 1979.

Gross beta, readings ran from 1 to 30 pCi/liter. The tritium readout showed the fallout to be a steady 450 pCi/liter. The flu symptoms and those of radiation poisoning are identical! Could it be another snow job or just a coincidence?

TABLE 4—*Continued*

Collection Period	*Concentrations expressed as pCi/liter* Gross Alpha	*Gross Beta*	*Tritium*
October 27, 1977-February 1, 1978	1	10	<450
February 1-April 25, 1978	1	14	<450
April 25-August 8, 1978	<2	9	<450
August 8-November 1, 1978	<1	5	<450
November 1, 1978-July 30, 1979	1	4	<450
July 30-September 25, 1979	<1	3	<450
September 25-November 16, 1979	<1	2	<450

TABLE 5

Radioactivity and Concentrations of Radionuclides in Rainwater

Location: La Grande at EOSC (northeast Oregon)

Collection Period	*Concentrations expressed as pCi/liter* Gross Alpha	*Gross Beta*	*Tritium*
August, 1971	<1	50	
September, 1971	<1	18	
October, 1971	<1	37	
November, 1971	<1	26	
February, 1972	<1	32	
March, 1972	<1	17	
August-September, 1972	<1	17	
December, 1972	<1	9	
January, 1973	<1	6	
February, 1973	<1	7	
April, 1973	<1	14	
May, 1973	2	6	
June, 1973	<1	1	
July, 1973	2	60	

TABLE 5—*Continued*

	Concentrations expressed as pCi/liter		
Collection Period	*Gross Alpha*	*Gross Beta*	*Tritium*
August, 1973	<1	9	
September, 1973	<1	3	
October, 1973	<1	2	
November, 1973	<1	2	
December, 1973	<1	3	
January, 1974	4	3	
February, 1974	<2	5	
March, 1974	<1	6	
April, 1974	<1	18	
May, 1974	1	21	
June, 1974	4	11	
July, 1974	1	25	
August, 1974	<1	70	
September, 1974	1	38	
October, 1974	1	23	
November, 1974	<1	8	
December, 1974	<1	7	
January, 1975	1	4	
February, 1975	1	9	
March, 1975	1	8	
April, 1975	<1	11	
May, 1975	<1	7	
June, 1975	<1	8	
July, 1975	<1	12	
August, 1975	<1	3	
September, 1975	—	60	
October, 1975	<1	4	
November, 1975	3	14	
December, 1975	<2	1	<450
January, 1976	—	—	<450
February, 1976	2	2	
March, 1976	<2	2	
April, 1976	<2	1	
May, 1976	<1	3	
June, 1976	—	2	

TABLE 5—*Continued*

Collection Period	*Gross Alpha*	*Gross Beta*	*Tritium*
	Concentrations expressed as pCi/liter		
July-August, 1976	—	2	
September, 1976	—	21	
October, 1976	5	18	<450
November, 1976	1	22	
February, 1977	<1	2	
March, 1977	<1	6	
April, 1977	<1	21	
May, 1977	—	19	<450
June, 1977	—	47	530
July, 1977	1	79	
August, 1977	<1	17	
September, 1977	<1	67	<450
October, 1977	<1	16	
November, 1977	<1	16	<450
December, 1977	<1	5	<450
January, 1978	<1	9	<450
March 1-16, 1978	<1	11	
March 16-31, 1978	<1	19	<450
April, 1978	<1	6	<450
May, 1978	<1	16	
June, 1978	<1	25	
July, 1978	<1	14	
August, 1978	<1	7	
September, 1978	<1	4	
November, 1978	<1	5	<450
December, 1978	<2	6	<450
January, 1979	<1	8	
February, 1979	<1	3	
March, 1979	<1	3	
April, 1979	<1	5	
May, 1979	<1	5	
June, 1979	<1	7	
July, 1979	1	31	
August, 1979	<1	5	
September, 1979	3	14	
October, 1979	<1	2	

TABLE 6

Radioactivity and Concentrations of Radionuclides in Rainwater

Location: Astoria

Collection Period	Gross Alpha	Gross Beta	Tritium
	Concentrations expressed as pCi/liter		
November 17, 1972-March, 1973	<1	4	—
March-June 27, 1973	<1	7	1000
June 27-December 12, 1973	<1	2	<450
December 12, 1973-February 12, 1974	<1	6	—
February 12-June 5, 1974	<1	18	<450
June 5-August 28, 1974	<1	16	<450
August 28-December 17, 1974	<1	9	<450
December 17, 1974-February 20, 1975	2	7	<450
February 20-June 19, 1975	<1	11	<450
June 19-August 25, 1975	2	7	<450
August 25-December 18, 1975	<2	2	<450
December 18, 1975-February 26, 1976	<2	3	<450
February 26-July 27, 1976	3	4	<450
July 27-December 7, 1976	1	6	<450
December 7, 1976-May 26, 1977	<2	6	<450
May 26-November 17, 1977	1	14	<450
November 17, 1977-March 7, 1978	<1	8	<450
March 7-June 1, 1978	<1	21	<450
June 1-September 13, 1978	<2	24	<450
September 13-December 13, 1978	<1	4	<450
December 13, 1978-March 8, 1979	<1	5	<450
March 8-September 18, 1979	<1	3	<450
September 18-December 6, 1979	<1	1	<450

TABLE 7

Radioactivity and Concentrations of Radionuclides in Rainwater

Location: Tillamook Bay

	Concentrations expressed as pCi/liter		
Collection Period	*Gross Alpha*	*Gross Beta*	*Tritium*
June 27-September 18, 1973	<1	8	<450
September 18-December 13, 1973	<1	4	<450
December 13, 1973-February 13, 1974	<1	6	—
February 13-March 26, 1974	<1	13	<450
March 26-June 6, 1974	<1	21	<450
June 6-August 27, 1974	<1	13	<450
August 27-December 18, 1974	<1	9	<450
December 18, 1974-February 21, 1975	<1	11	<450
February 21-June 20, 1975	2	11	<450
June 20-August 26, 1975	1	5	<450
August 26-December 18, 1975	<2	3	<450
December 18, 1975-February 26, 1976	<2	4	<450
February 26-July 28, 1976	<1	3	<450
July 28-December 8, 1976	<1	8	<450
December 8, 1976-March 25, 1977	<1	8	<450
March 25-May 27, 1977	<2	17	<450
May 27-November 18, 1977	<1	16	<450
November 18, 1977-March 8, 1978	<2	10	<450
March 8-June 6, 1978	<1	16	<450
June 6-September 14, 1978	<1	7	<450
September 14-December 14, 1978	<1	4	<450
December 14, 1978-September 19, 1979	<1	2	<450
September 19-December 7, 1979	<1	1	<450

TABLE 8

Radioactivity and Concentrations of Radionuclides in Rainwater

Location: Dayville (northcentral Oregon)

Collection Period	*Gross Alpha* (Concentrations expressed as pCi/liter)	*Gross Beta*	*Tritium*
January 25-April 25, 1974	—	—	<450
April 25-July 21, 1974	2	41	<450
July 21-September 12, 1974	—	—	<450
September 12, 1974-February 16, 1975	<1	7	<450
February 16-March 18, 1975	<1	16	<450
March 18-November 9, 1975	<1	12	<450
November 9, 1975-January 20, 1976	3	3	<450
January 20-April 12, 1976	—	17	<450
April 12-July 30, 1976	<1	5	<450
July 30-November 13, 1976	<1	3	<450
November 13, 1976-January, 1977	—	—	<450
January-May 30, 1977	<2	25	<450
May 30-November 12, 1977	1	16	<450
November 12, 1977-January 12, 1978	1	12	<450
January 12-May 28, 1978	2	19	<450
May 28-November 25, 1978	1	14	<450
November 25, 1978-October 14, 1979	1	5	<450

Some new fission products radionuclides, associated with the Chinese fallout are Zirconium-Niobium 95, Cesium 137, Ruthenium 103, Ruthenium 106, and Cerium 141 and 144. The occurrence of these radionuclides in the sampling media was attributed to the atmospheric weapons testing source. However, since they were atmospheric tests and not surface detonations, it is possible that products or ingredients were deliberately incorporated in the warhead to determine what the reaction of the American (guinea pigs) people would be to that type of radiation! Atmospheric testing has got to stop!

A war is something that goes on from sixty days to years. There is not and there never will be such a thing as an *atomic war!* There will be utter destruction of civilization, and all life as we know it will be over in *thirty minutes!* Evacuation couldn't even get started. The military-industrial complex will cease to exist, and probably so will you! Perhaps I went too far when I said there never was an atomic war? Look at Mars, Jupiter, and any other planet where there should be life? Will God allow humanoid history to repeat?

Could it be that many of the so-called flu epidemics are really the effect of radioactive particles lodged in the lungs, liver, or some other part of the body? Oregon is not alone with fallout problems. Many other states have heavy concentrations of fallout due to rainout, snowout, or just because huge clouds of debris are low as they pass over.

Doctors may tell you when you make an office visit that you are suffering from a virus of some sort and that it's all over town. (I wonder if a test for an irradiated organ was ever made in an autopsy on a person who supposedly died of flu.) I went that route and had my flu shot. If I make it another twenty years I'll thank God! However, doctors prescribe for a cough and sore throat and, surprisingly, that gives some relief!

The old cliché that a cold will take a week to cure no matter what you do doesn't apply here. When your throat hurts and your voice changes, you feel like hell for two or three weeks, but you don't ache all over, lose weight, or get weaker. If it doesn't do you in, it will go away for some time, perhaps several days or weeks, but it comes back. The second time around may not be so long or as bad. Perhaps the body can fight it off. However, I don't think we need to worry about birth control or overpopulation for a long, long time!

The National Center for Disease Control in March, 1980, stated that influenza outbreaks were blamed for driving flu and pneumonia deaths in the United States to 4,450 since the first of the year—1,000 more than expected statistically. Oregon was one of ten states reporting epidemic levels of flu.

After the outbreak of influenza in Oregon ran for two weeks, it showed signs of tapering off. Flu fatalities persisted, however, remaining above the so-called epidemic threshold for seven consecutive weeks!

One fellow received the largest internal dose of radioactivity ever known and he's still alive five years later, so don't give up the ship, folks. Perhaps *they* can find a way to clean out the upper air and turn the good earth back to the healthy orb it once was.

Doctor should test for radiation poisoning and learn the method used to capture plutonium particles that concentrate in the liver. *There is a way* (see page 146). Atoms are captured and passed out through the kidneys, which do not seem to be affected by radiation but can be poisoned by the chemical effects!

One study shows that fallout from atomic bombs tested twenty years ago is responsible for more than 116,000 cases of lung cancer in the United States alone.

In fourteen years, from 1950 to 1963, 758 people in Oregon died of liver cancer; from 1964 to 1979, a period of sixteen years, 412 people died of it, a total of 1,170 deaths—definitely not a normal situation.

The following tables show the breakdown of the number of deaths from liver cancer according to county residence in Oregon from 1950 to 1979, and a breakdown of liver deaths of Oregon residents according to sex, age, and year from 1962 to 1979.

It is possible to read many things from these tables. A summary shows that men are much more susceptible to liver cancer than women. For those who would deny radiation as the killer, perhaps they would believe that it is just an epidemic of wild cells going on a rampage all over the world. If so, what is causing it?

The main damage to mankind and his environment occurred from 1945 to 1962, the year the big fellows ceased and desisted from detonating atmospheric fission and fusion bombs.

Cancer is a slow grower. The tables show the steady increase over those seventeen years in the incidence of liver cancer. One reassuring fact is that if you reach the age of ninety, you've got it made.

RAINWATER AND SNOWOUT SUMMARY

Radiation debris clouds spread out like a huge pancake. Depending on the vagaries of the upper air winds, gravity, and the detonation of other bombs in the atmosphere, they may merge, if they haven't already, into a single, hot, radioactive shell completely around the globe! That stuff will be up there for years.

Portland, Oregon, has a thirteen-year project for drilling thirty water wells at an estimated cost of twenty-three million dollars. This is cheap life insurance for the lives of hundreds of thousands of people! Mayor Frank Ivancie, a true representative of the people, presented the idea to the City Council in 1974. He, as commissioner of Public Works, was well aware of the weak link in our life support system. There seems to be aquifers (large underground lakes) spotted all over the world. Some are in desert regions where the rain just disappears into the sand! Thank God for a good roof over these natural reservoirs of water. We better get more drillers working PDQ. Fallout is still coming down all over the world!

There is always the threat of volcanic ash fallout, earthquakes, vandals, and terrorists. Vandals and terrorists, we can exterminate! Then there will be no prisoners for other terrorists to free by more acts of terrorism! Acts of Mother Nature we don't have to like, and we can defend ourselves against some. Acts of God we may not like, but God knows best!

It has been suggested that the detectable level of tritium be listed at 450 pCi/liter, as it was almost always below that. According to the data above, it seemed to hover around 300 to 350 pCi/liter, which was harmless to the skin. At times it did increase much above that. Whatever the reason is, logically, the data show a cumulative fallout, a constant buildup over land and water.

TABLE 9

Primary Carcinoma of Liver Deaths by County of Residence, Oregon, 1950-1963

County of Residence	1950	1951	1952	1953	1954	1955	1956	1957	1958	1959	1960	1961	1962	1963
Baker	—	—	1	—	—	1	1	—	1	—	1	—	1	3
Benton	1	—	—	1	2	—	—	—	1	2	2	—	1	2
Clackamas	1	1	3	3	3	1	2	2	3	5	2	5	5	5
Clatsop	1	1	—	—	—	1	—	2	1	1	—	—	1	—
Columbia	1	—	1	1	—	1	1	—	—	—	—	—	—	1
Coos	—	1	2	—	1	2	—	—	1	1	1	1	—	1
Crook	—	—	—	—	1	—	—	—	—	1	—	—	—	1
Curry	—	—	—	—	—	—	1	—	—	—	1	—	1	—
Deschutes	1	—	—	1	1	—	—	—	1	—	1	1	3	1
Douglas	2	—	1	3	1	1	2	3	1	3	2	1	2	—
Gilliam	—	—	—	—	—	—	1	—	—	—	—	—	—	—
Grant	—	—	—	—	—	—	—	—	—	1	—	—	—	—
Harney	—	—	—	—	—	—	—	1	—	—	—	—	—	—
Hood River	—	—	—	—	—	—	1	—	2	—	—	—	2	—
Jackson	—	1	3	6	3	1	2	1	1	5	2	6	1	2
Jefferson	—	—	—	—	—	—	1	—	—	—	—	—	1	—
Josephine	—	1	—	1	—	1	—	1	—	—	2	2	1	1
Klamath	1	1	2	—	3	1	2	1	—	1	3	1	4	2
Lake	—	—	—	—	—	—	—	—	—	—	1	—	2	—
Lane	2	1	2	1	4	2	3	5	3	9	8	5	3	4

Lincoln	—	2	—	1	—	1	—	2	—	1	—	—	—	1
Linn	1	1	—	2	—	3	2	3	2	—	2	2	2	1
Malheur	—	—	1	1	—	—	3	—	1	1	—	3	—	2
Marion	3	2	4	2	3	3	2	2	3	2	7	7	3	9
Morrow	—	—	—	—	—	1	—	—	—	—	—	—	2	—
Multnomah	18	16	15	20	12	24	22	29	14	21	22	33	21	26
Polk	—	—	2	2	—	1	—	1	—	—	2	—	1	—
Sherman	—	—	—	—	1	1	—	—	—	—	—	—	—	—
Tillamook	1	1	—	1	1	—	1	—	—	—	1	2	3	—
Umatilla	3	1	2	3	1	2	2	1	1	2	2	3	2	1
Union	1	1	—	—	1	1	2	1	1	1	2	—	—	1
Wallowa	—	—	—	1	2	1	—	—	—	—	—	—	—	—
Wasco	—	1	—	1	1	—	1	—	1	—	—	—	—	—
Washington	—	3	—	—	3	2	—	1	2	2	4	1	2	2
Wheeler	—	—	—	—	—	2	—	—	2	—	—	—	—	—
Yamhill	—	—	2	—	3	—	2	3	—	1	1	2	4	—
Total	37	35	41	51	47	54	54	59	42	60	69	75	68	66

Source: Oregon State Health Division, Vital Statistics Section, Portland, Oregon.

TABLE 10

Primary Carcinoma of Liver Deaths by County of Residence, Oregon, 1964-1979

County of Residence	1964	1965	1966	1967	1968	1969	1970	1971	1972	1973	1974	1975	1976	1977	1978	1979
Baker	—	—	1	—	—	1	—	—	—	—	1	—	1	1	—	—
Benton	—	—	2	—	—	1	—	1	—	—	—	1	—	—	—	—
Clackamas	1	—	—	1	2	—	3	2	2	1	—	2	1	1	1	1
Clatsop	—	—	—	—	—	2	—	1	—	—	—	—	1	1	—	—
Columbia	—	—	2	—	—	—	—	—	—	—	—	—	—	—	—	—
Coos	2	—	2	—	—	2	2	3	4	1	—	2	1	—	1	—
Crook	—	—	—	—	—	—	—	—	1	1	—	—	—	—	—	—
Curry	—	—	—	—	—	—	—	—	—	—	—	—	—	—	—	—
Deschutes	1	—	—	—	—	—	—	—	—	1	—	3	1	—	—	—
Douglas	2	—	1	—	—	1	3	1	—	4	2	—	1	2	4	1
Gilliam	—	—	—	—	—	—	—	—	—	—	—	—	—	—	—	—
Grant	—	—	—	1	1	—	—	—	—	—	—	—	—	—	—	—
Harney	—	—	—	1	—	—	—	—	—	—	—	—	—	—	—	—
Hood River	—	—	—	—	—	—	—	—	—	—	—	—	1	—	—	—
Jackson	3	—	—	—	1	2	1	5	3	3	—	2	—	1	1	2
Jefferson	—	—	—	—	—	1	—	—	—	—	—	—	—	—	—	—
Josephine	2	—	1	—	—	—	—	—	1	1	—	1	2	—	—	1
Klamath	—	—	—	—	—	—	—	—	—	1	1	2	—	2	1	1
Lake	—	—	—	—	—	—	—	—	—	—	—	—	—	—	—	—
Lane	1	2	2	3	3	3	2	4	—	2	3	2	1	—	2	5

Lincoln	—	—	—	—	—	—	—	—	—	—	—	—	—	2	—	3
Linn	—	1	—	1	1	3	1	2	—	1	1	2	3	2	3	2
Malheur	—	—	—	—	—	—	1	—	—	—	—	—	—	1	—	—
Marion	5	—	2	1	3	4	—	2	1	—	3	2	3	2	—	—
Morrow	—	—	—	—	—	1	—	—	1	—	—	—	—	—	—	—
Multnomah	9	6	6	12	11	5	7	9	6	7	6	11	12	4	5	7
Polk	—	—	—	—	—	—	1	—	1	—	1	1	2	2	—	—
Sherman	—	—	—	—	—	—	—	—	—	—	—	—	—	—	—	—
Tillamook	—	—	—	—	—	—	—	—	—	—	—	—	—	1	3	1
Umatilla	—	1	—	1	—	1	1	—	—	—	2	1	—	1	2	—
Union	—	—	—	—	—	—	—	1	—	1	—	—	—	—	—	—
Wallowa	1	—	—	—	1	—	1	—	—	—	—	—	—	—	—	1
Wasco	—	—	1	—	—	1	—	—	—	—	—	—	1	—	—	—
Washington	—	2	—	1	3	1	2	1	1	2	2	2	2	—	3	2
Wheeler	—	—	—	—	—	—	—	—	—	—	—	—	—	—	—	—
Yamhill	1	1	—	—	1	—	—	—	1	—	—	—	1	1	1	—
Total	28	13	20	22	27	29	25	32	22	26	22	34	34	24	27	27

Source: Oregon State Health Division, Vital Statistics Section, Portland, Oregon.

TABLE 11

Number of Deaths Due to Liver Cancer, by Sex, Age and Year, 1962-1979, Oregon Residents

Total	1	3	1	1	2	3	3	1	5	13	19	30	47	74	94	68	47	48	28	5	493
Age	1 Yr.	1-4	5-9	10-14	15-19	20-24	25-29	30-34	35-39	40-44	45-49	50-54	55-59	60-64	65-69	70-74	75-79	80-84	85-89	90+	
1979																					
M										1	1	1	1	3	6	3	1	3			20
F									1	1		1	1	1	2	1		1	1		10
1978																					
M	1	1					1			1			2	4	4	2	1	2	1		20
F							1						1	1	4	1		3	3		14
1977																					
M							1			1	1				4	2	4	2			15
F				1							1	2		1	3	3	3			1	15
1976																					
M		2									1	1	2	2	1	4	4	2	2	1	22
F					1						1	3		3	2	1	1	2	1		15
1975																					
M									1		1	1	3	3	4	4	3	1			21
F										1		3	1		1	2	1	3	3		15
1974																					
M										2		3	2	6	2	2	1	1			19
F										1			1	3	1	1	2				9
1973																					
M													1	2	5	2	3	1	1		15
F											1		2	2	2	1	1	1		1	11
1972																					
M										1	1		1	2	1	2		3	1		12
F									1				2	2	3	3		2			13

1971																					
M												1	3	4	7		1	2		1	19
F													2	4	2	3	1	2	2	1	17
1970																					
M					1								1	1	1	2	3	3	5		17
F												2	1		1	1	2		2		9
1969																					
M						1					1	3	1	6	3	3		2			20
F											2			1	2	1		3	1		10
1968																					
M											2		2	6	3	3		1			17
F												2	1	3	1	1	1	1	2		12
1967																					
M									1		1		1		5	2	1	1			12
F											1	1	3	1		2	2				10
1966																					
M						1							1	2	6	2	1	1			14
F													2		2	2					6
1965																					
M													2	2	3	3	1				11
F														1					1		2
1964																					
M									1	1		1	3		5		3	2			16
F						1							1	1	1	4	2	1	1		12
1963																					
M								1		1	2	2	3		3	1	1	1			15
F												1			2	3	2	1	1		10
1962																					
M			1							1	1	2		6							11
F										1	1			1	2	1	1				7
Total M	1	3	1	0	1	2	2	1	3	9	12	15	29	49	63	37	28	28	10	2	296
Total F	0	0	0	1	1	1	1	0	2	4	7	15	18	25	31	31	19	20	18	3	197

Source: Oregon State Health Division Vital Statistics Section, Portland, Oregon.

3

More Atomic and Thermonuclear Atmospheric Bomb Tests

HOW *THEY* BLEW OUR WEATHER

I refer you to the cover photo as one of the causes of our freakish weather. The People's Republic of China, hereafter referred to as China, began testing in October, 1964, at their Lop Nor test site in southwest China. This was not too long after we allowed a nuclear physicist, a Chinese trained by us, to escape to China! On September 26, 1976, China exploded its *nineteenth* atmospheric atom bomb. This was only a 20/200 KT device equal to 200,000 tons of TNT! Rated small as atom bombs go, it was a low altitude (45,000 feet), low yield bomb, detonated in the troposphere, the part we live in!

Constant surveillance of radioactivity levels in the United States is maintained through EPA's Environmental Radiation Ambient Monitoring Systems (ERAMS). These systems were all activated, with the twenty-one continuously operating stations and forty-six standby stations located in the United States, Puerto Rico, and the Canal Zone, as well as in all the free world nations. It's too bad we can't obtain the results from the rest of the free world nations. I have tried but received no reply! Daily samples are taken of air particulate fallout. See Fig. 16—Source: EPA-520/5-77-002, assessment of fallout in the United States from atmospheric nuclear testing on September 26, 1976, by the People's Republic of China. This shows the path of travel for the debris cloud as it traveled from the China test site. Note the travel times on the route lines.

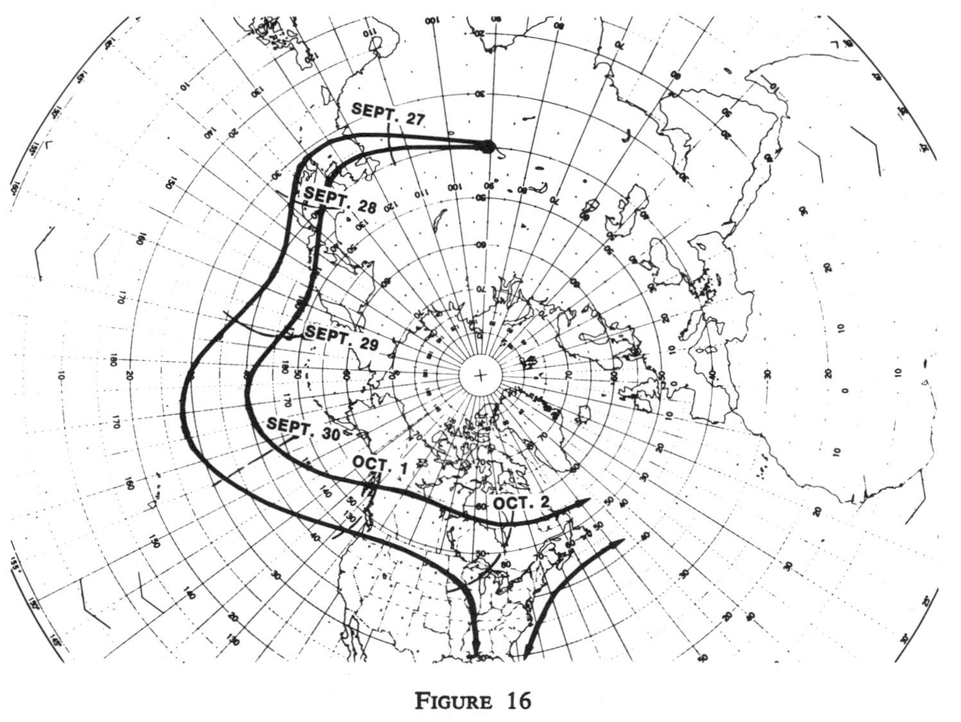

FIGURE 16

Analysis of path of debris, at approximately 30,000 feet, from the Chinese nuclear detonation of September 26, 1976. *(Courtesy Environmental Protection Agency [EPA])*

On September 30, 1976, the leading edge of the first debris cloud reached the West Coast (four days). Traveling high and fast in the prevailing jet streams' easterly flow, it arrived in British Columbia and over the northwestern United States about 1:00 P.M., Pacific standard time. The first confirmation came from air samples taken at a station in Richland, Washington. The midwestern and eastern stations tracked the ever-expanding cloud over South Dakota, Missouri, Mississippi, and Florida. It then turned northward along the East Coast (see Fig. 16). The southern portion of this cloud mass passed over the northern portions of Washington, Idaho, Montana, North Dakota, and Minnesota on October 1, 1976. On the night of October 1, a low pressure center formed over Illinois, Indiana, and Ohio and caused a severe atmospheric disturbance that intersected the southern portions of the cloud. The radioactive particulates interacted as heavy cloud-seeding, and heavier than normal precipitation occurred, bringing down radioactive particles all over northeastern Maryland, southwestern Pennsylvania, Delaware, New Jersey, southwestern New York, western Connecticut, and western Massachusetts. Many radioactive particulates were detected, not in amounts requiring protective action but in amounts sufficient to do a bang-up, cloud-seeding job! Radiation readings were extremely small but were building up. Radioactivity was highest on the eastern seaboard the first ten days of October. The highest was recorded in the Deep South on October 18 to 20. This was from the cloud's second time around the world! The first time the cloud was much too high.

The highest amount of iodine-131[i] was found in pasteurized milk at Baltimore, tested at 155 pCu/l (pico-curies per liter), though several agencies registered hot spots of 1,000 pCu/l, compared to 750 rems dose, which is estimated to be lethal, depending on how long the exposure is. This was from individual dairies and was no doubt due to the "fountain effect." As a protective measure, some dairy herds were ordered off pasture land and fed stored feed. Dr. William D. Rowe, EPA assistant administrator for radiation programs, indicated that even the highest levels were well below where protective action was needed. However, a second cloud from that same Chinese test was traveling over the Pacific at a lower altitude of 20,000 feet. Being slower and lower, it took nine days (see Fig. 17). Match your bad weather with this one! The clash of cold air with warm air, plus continuous interaction with cloud seeding, didn't help things a bit. Drifting

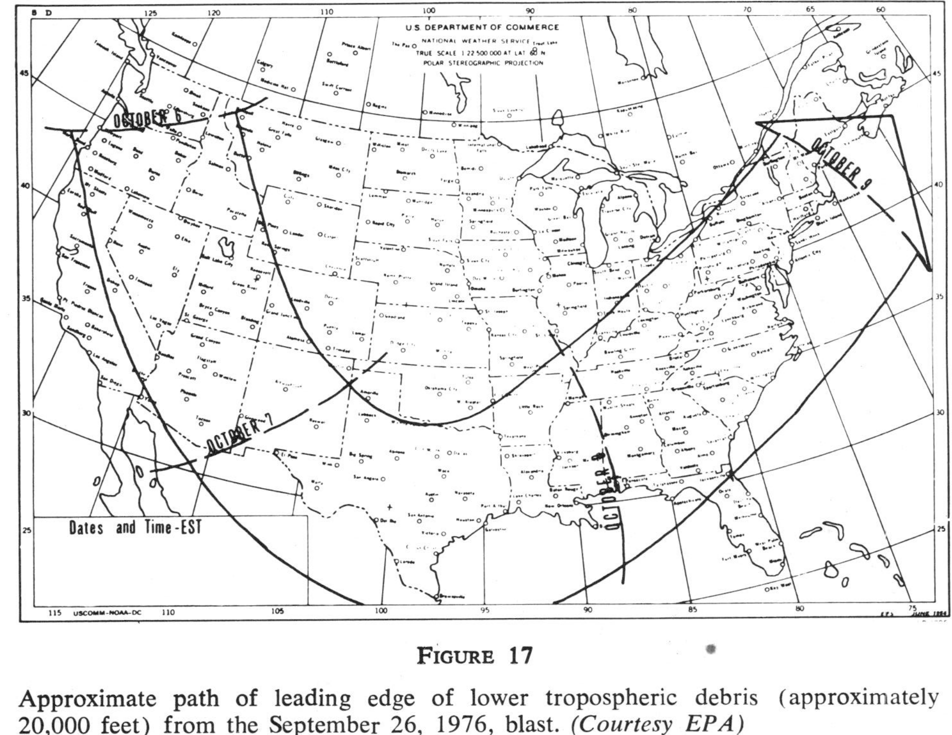

FIGURE 17

Approximate path of leading edge of lower tropospheric debris (approximately 20,000 feet) from the September 26, 1976, blast. *(Courtesy EPA)*

FIGURE 18

Downstream view of the normally dry Los Angeles River. *(Courtesy Los Angeles Flood Control District)*

FIGURE 19

Gushing waters of the Los Angeles River during the storm. (Los Angeles Times *photo*) south over the Pacific, it arrived over the California coast on October 6, 1976. It then moved eastward in a *U*, or horseshoe shape, across the United States and up the East Coast.

This cloud was 400 miles wide and heavy with radioactive particulates. This nineteenth test by China placed most of the radioactive debris in the troposphere and, moving slower, accounted for the heavy fallout and hot spots over the southern states. In California, state officials, having had a full-scale drought for months, were determined to get all the moisture from the storm front that was passing over. Fig. 18 is a downstream view of the normally dry Los Angeles River. It was already raining heavily. In a desperate attempt to get it all (and they did), they conducted cloud-seeding on their own! Evidently they were not aware of or had not been told that the radioactive particulates were already seeding the clouds. They did this under protest, as the downpour was far heavier than usual. The seeding brought on a cloudburst. Millions in damage resulted and many civil suits were filed against the state. Fig. 19 shows water gushing down the normally dry bed of the Los Angeles River at Los Feliz Boulevard.

As with all others, the radioactive fallout from this cloud was quiet and couldn't be heard on tin roofs. In snow or rain, the fallout is invisible but deadly, silent, poisonous. The most direct route to man, next to skin contact, is the forage cow route, as iodine-131[i]. Human metabolism concentrates all ingested iodine in the thyroid glands; these, in turn, affect endocrine glands. The buildup is cumulative. In time, deaths from thyroid cancer and leukemia will increase! It's bad enough now. How fast it will get worse, no one knows! Radioactive fallout or radiation exposure can be the unknown cause of many problems for mankind. Now, so much for the *little bomb.* Let's take a look at one *twenty times larger!*

NOVEMBER 17, 1976

On this date, China sent up into the stratosphere a 4/MT bomb—equal to *4 million tons of TNT!* It took five days for the little bomb's radioactive cloud to reach the United States. It took three days for the big one, this one spreading its own poison and that of the smaller one higher, and wider, than before, via the jet streams, because it was exploded smack dab in the center of the returning cloud of September 26, 1976!

Being detonated in the stratosphere, creates a fountain effect, hurling air heated to sunlike temperatures far out into space, rising higher and higher only to come crashing down, at another time and another place, extremely cold and far removed from where it happened, as the world turns, and you *wonder* what's causing our freak weather? Was this a ruse by China to deploy radiation over the United States?

This bomb of November 17, 1976, was predicted to have a much wider north and south dispersion and was far, far, larger than predicted. Uncle knows and meteorologists know what fouled up Chicago's weather! A local newspaper reported that a large grant was given to a group to study whether the aircraft over O'Hare Airport was the cause of so much freak weather there. Please pardon the pun, was it another snow job by Uncle? Was it more money spent for a cover-up? Was it seven feet of *hot* snow? Did the word go out—"Don't touch it, it's radioactive"? Did gravity finally get a good hold and bring it down? And to think that is why the mayor lost his job. He didn't remove the snow!

It took this huge cloud one day to cross the United States (see Fig. 20) after arriving on the West Coast November 20.

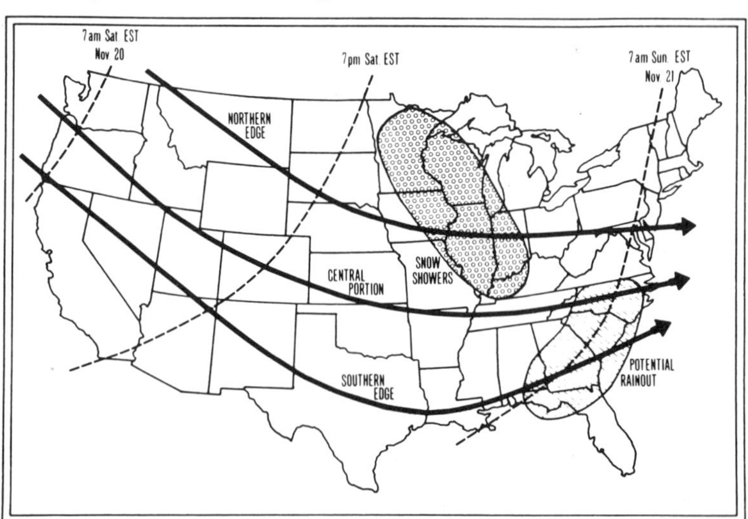

FIGURE 20

Predicted movement of air mass containing radioactive debris across the United States and possible areas of rainout from the air mass following the Chinese nuclear detonation of November 17, 1976. *(Courtesy EPA)*

It passed southeasterly over the states and out to sea by November 21. This was the last was my sentiment at the time this was written, but it wasn't. *They,* not just China, have seeded the world with *poisonous,* radioactive particulates. *They* don't seem to know that the area of the heaviest fallout is the place of origin, except for a land burst. The first time around, the cloud is much too high but by the time it gets back to where it started from, fallout will be coming down real good! Of course, the rest of the world will get it also!

During the 1967-1971 period, there were many intrusions of fresh fission product activity into the Columbia River environment from Chinese atmospheric nuclear tests. A summary of these detonations for the years 1964-1977 with their approximate yields in kilotons (Table 12) follows. Besides the Chinese atmospheric detonations, underground detonations by the United States on the Nevada test site and by the Soviet Union, primarily at their Semipalatinsk and Novaya Zemlya testing areas, took place. However, these underground tests were considered to have contributed only negligible radioactivity to the atmosphere.

The fission product radionuclides associated with fallout are: zirconium-niobium-95, cesium-137, ruthenium-103, ruthenium-106, and cerium-141 and 144. The occurrence of these radionuclides in the sampling media was attributed to the atmospheric weapons testing source.

China has detonated three more bombs since No. 20. I hate to think that the day may come when many or perhaps all of the so-called third world countries may develop atomic weapons. Woe be to them who furnishes the enriched materials. It could be a boomerang! These countries are the children of the world. They want everything Mother and Dad and Big Brother have, and they want it right now—much like our own children at home or just after they leave home, except that they seldom turn on us!

The United Nations has a big job ahead of it. Like OPEC, third world nations, to get what they want or to avenge themselves for past maltreatment, would not hesitate to blackmail the adult nations. But we cannot expect any help from the UN on matters of atomic fallout when it cannot even pressure Mexico to stop polluting the Coatzacoalcos River and the Gulf of Mexico. With over 300,000 people concerned, it may soon make Japan's mercury incident at Minamata Bay look like a drop in the bucket in comparison. If Mexico stopped right now, it would be over a century before the damage would be undone! Talk about

TABLE 12

Concentrations of Radioactive Debris from Chinese Atmospheric Nuclear Tests 1964-1977

		Approximate Yield (Kilotons)	
Year	*Date*	*Total*	*Fission*
1964	Oct. 16	20	20
1965	May 14	40	40
1966	May 9	300	300
	Oct. 28	20	20
	Dec. 28	300	300
1967	June 17	3000	1600
	Dec. 24	20	20
1968	Dec. 28	3000	1900
1969	Sep. 29	3000	1800
1970	Oct. 14	3000	1500
1971	Nov. 18	20	20
1972	Jan. 7	<20	<20
	Mar. 18	20-200	20-200
1973	Jun. 27	2000-3000	1400
1974	Jun. 17	200-1000	200-1000
1976	Jan. 23	<20	<20
	Sep. 26	20-200	20-200
	Nov. 17	4000	2000
1977	Sep. 17	<20	<20

rotten human relations! It's just as deadly as radioactive fallout, which they are receiving along with the rest of us, worldwide, plus their visible pollution!

It's obvious that Mexico is dumping millions in heavy metals into their river, bay, and the Gulf, when they should be salvaging it. They are also burning off gas at the wellhead. This creates hundreds of minivolcanoes that blow the atmospheric balance as St. Helens did for the West Coast! That gas could be liquified and sold all over the world or piped to the United States. It also blows our weather along with the tropical jet stream. Those gas jets could produce a lot of first-class hurricanes. Read and pray!

THE BIG, HOT BLANKET

There has been considerable apprehension about aircraft flying through the heavily radioactive clouds. A slow cook was feared, which might show up after a period of years or instantly, depending on exposure time.

As expected, there were no real problems at normal commercial aircraft altitudes (up to 40,000 feet). Measurements made aboard aircraft indicated that exposures from radioactive materials at altitudes of 30,000 to 35,000 feet would be increased by about only 2 percent over the exposures normally received at these altitudes from cosmic radiation. Exposure at lower altitudes was even smaller. The slightly increased exposure due to suspended fallout debris was roughly the equivalent of increased cosmic radiation when flying at 32,000 feet as compared to 30,000 feet.

EPA consulted with the FAA (Federal Aviation Agency), ERDA, and the Air Force in assessing the impact of airborne radioactive materials on aviation. All of the agencies agreed that there would be no problems with passenger exposure at normal altitudes. Therefore, no recommendations were made to divert flights around the path of clouds carrying fallout debris. EPA advised that business should continue as usual for all regular jet air travel.

However the Anglo-French Concorde SST, flying at 50,000 feet, presented a problem. The energy from the November 17, 1976, bomb pushed almost 100 percent of the debris cloud into the stratosphere! At this altitude, atmospheric diffusion would spread it over the entire globe, with the maximum surface deposition occurring over the middle latitudes of the hemisphere, where the explosion originated.

The SST flight pattern is much different from the regular behavior of commercial aircraft. Typically, the Concorde climbs quickly to about ten miles. As fuel is consumed, it maintains a slow climb to about twelve miles (63,360 feet) shortly before descent for landing. At these flight levels, the SST would be much closer to the debris cloud center than would U.S. civil aircraft (see Fig. 21).

Of course, there may come a day when the atmosphere reaches a saturation point for radioactive particulates and the detonation of just one more bomb will cause complete precipita-

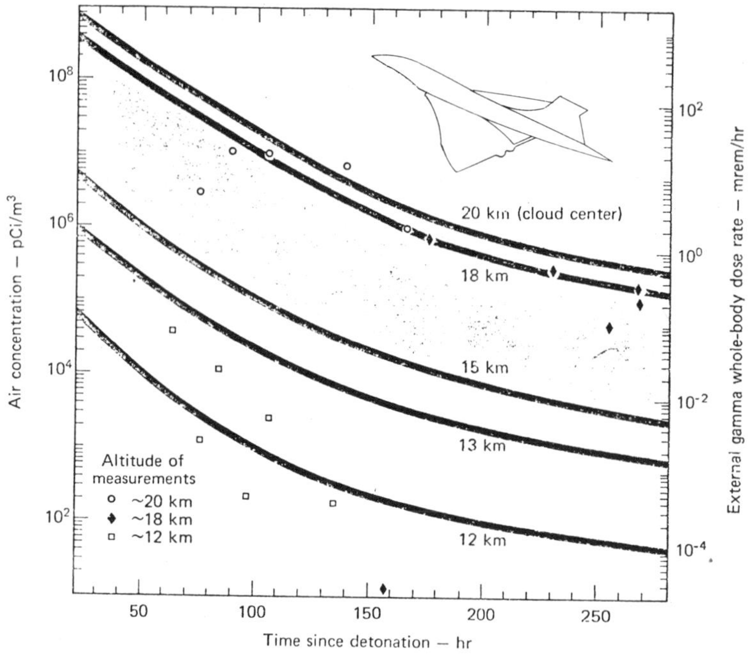

FIGURE 21

Predicted (curves) and observed (data points) air concentrations and dose rate for the debris cloud from the November 17, 1976, Chinese detonation. The illustration shows the typical flight level for the Concorde SST. *(Courtesy National Oceanic and Atmospheric Administration [NOAA])*

tion or disrupt the Van Allen radiation belts to such a degree that cosmic radiation from the solar winds that put them there in the first place will burst through whatever opening is made and alter the magnetic field in the atmosphere enough to pull down and clean out completely the air ocean above the earth's surface! If what's up there now should suddenly be pulled down, or develop an affinity for aluminum, all high-altitude aircraft would become jumbo microwave ovens. Depending on exposure time, a person could become literally half-baked and not know it! How long do we have to see repeated disruptions of our weather with all kinds of extremes before we realize that it's atmospheric bombs that are blowing it! No government will accept or admit it, even though *in flagrante delicto* (caught in the act).

Actually, on a fly through the cloud center, it was determined that as long as you stayed inside the aircraft, there was no danger from radioactive exposure! There was an imposing array of devices to check and measure radioactivity from the ground up—a geiger counter, a proportional chamber, passive dosimeters, air sampling, and scintillation detectors. The gas proportional unit was packaged in a suitcase with a self-contained strip chart recorder. It was started before takeoff and provided a log of all radioactivity seen during the flight.

Fig. 22 shows the locations of smears taken. This was done with the thought that the maintenance crews might be threatened by radioactive particulates that might build up on the exterior SST surfaces. None was found in any dangerous amounts.

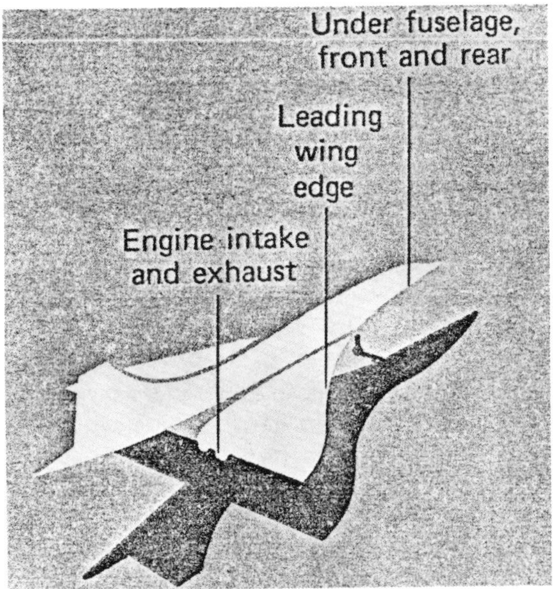

FIGURE 22

Designated surface-contamination swipe locations on the Concorde. *(Courtesy NOAA)*

Of course, there is our hydrogen bomb that the Chinese haven't developed yet and the neutron bomb that only kills people without destroying property. What it is and how it does it comes later. Then there is the cruise missile and others classified secret that we don't know about.

BOMBS OF 1977

It goes on and on. On September 17, 1977, at 3:00 A.M., EDT, The People's Republic of China set off another bomb, No. 21! This one was 20 kilotons. The Environmental Radiation Monitoring System stations (ERMS) were activated, supplementing the twenty-two that are continuously working. This is a very important part of our defense system allowed us in the SALT treaty!

Just keep your mind on the forage cow milk-to-man radiation path. That we can do something about, after it happens. What about the city water reservoirs open to the wild blue yonder? All our enemies have to do is sit back and wait for America to complete its slow suicide with the chemicals in our food and water, in addition to radioactive particulates. Why? When there is available filtration plants that will remove all bacteria and water-borne viruses, perhaps even neutralize radioactive particles at the point where the pipelines go underground. Some of our own people are doing their best to hurry our demise along with hard drugs! Let's regain the respect of the world!

Figs. 23 and 24 show estimated rainout on September 21 and 22, 1977. Rainwater samples of September 24, 1977, showed nuclear debris at Anchorage, Alaska; Portland, Oregon; Idaho Falls, Idaho; and many other cities. Cover your reservoirs and dig your wells for survival *now! The fire in the sky is already up there!* There is little likelihood of a nuclear war! There will be no winners! These bombs have come and gone, so have those of previous years. There hasn't been a map made of the potential debris cloud that wasn't underestimated! They don't just pass over to be forgotten. Every fourteen days they come back! They could even merge with others and be much more deadly. *God forbid a saturation point!*

The United States and the USSR *ceased* atmospheric testing in 1962, apparently knowing that *they* were fouling up the weather. ERDA monitors all nuclear explosions; these people at the Lawrence Livermore Laboratory, in Livermore, California, plus many other locations, in addition to state and private laboratories, are the real heroes of this deadly battlefield. They are digging out the facts as they find them. Though the facts are not always available to the public, nor are they always publicized, they are there to be found somewhere, somehow, if you persist.

With an estimated 380,000 thyroid cancers expected during

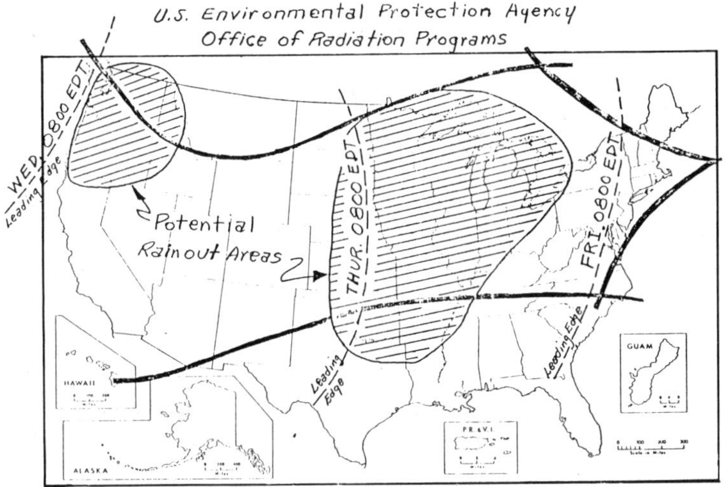

FIGURE 23

Predicted movement of air mass containing radioactive debris across the United States and areas of potential rainout, September 21, 1977. *(Courtesy EPA)*

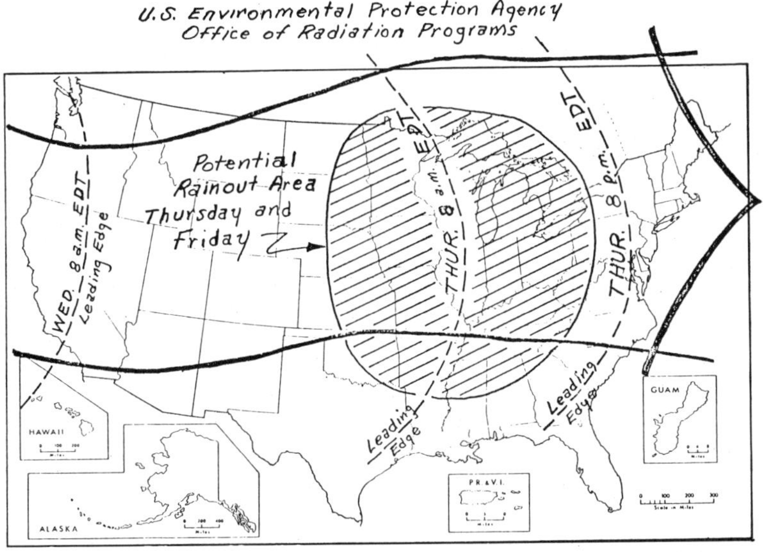

FIGURE 24

Predicted movement by NOAA of air mass containing radioactive debris and areas of potential rainout, September 22, 1977. *(Courtesy EPA)*

the next forty-five years, in addition to the thousands of cancers afflicting other organs of the body, what is it that is causing all this, along with so much leukemia? If it isn't radiation, is it something touted by HEW or the pharmaceutical companies? Laboratory research has confirmed that chlorine is one of the carcinogens. If it were not already in almost all of our drinking water, would HEW allow it to be used? Could it be fluoride in the drinking water, along with deadly radioactive particulates?

It takes about fifteen days for a radioactive cloud to circle the earth, depending on the altitude of the detonation. Around and round they go. If they are punched up into the stratosphere and if they stay up there for 45 years, the more dangerous ones, such as lanthamum-140, beryllium-7, strontium-89 and 90, cesium isotopes, and many others, will no longer be a threat to mankind. Of course, if they are pulled down by gravity sooner—well, no problem. Just stay indoors; their estimated life is now ninety years instead of the 240,000, as at first thought. Mankind and Mother Earth will get the full benefit of the radiation from the shell of radioactive particulates as they slowly decay the first 45 years!

BOMBS OF 1978

Here we go again! On March 14, 1978, about midnight, the People's Republic of China let go with another 20 kiloton fire cracker, No. 22!

Barbara Blu, of EPA, as deputy administrator, issued the notice in the *Environmental News* of Thursday, March 16, 1978, that the results of EPA's air and precipitation monitoring will be used to identify areas of rainfall contamination in order to alert state health agencies to monitor raw milk from farms in those areas. EPA's Pasteurized Milk Monitoring Network will then determine any exposure to the United States population that may occur from milk consumption.

Evidently, monitoring drinking water will be up to state health departments. See Tables 3-8 for the sampling of rainwater during the intrusion of radioactive debris clouds across Oregon. The Portland samples were taken on top of the State Office Building, ten stories up.

The United States has long opposed atmospheric testing of nuclear weapons. The USSR, France, and England also agreed to stop it. Others either don't give a damn or aren't aware of the plague of cancer they are spreading all over the world! If people take the trouble to distill their drinking water it will remove everything but tritium. All can be removed by ion exchange!

By March 25, 1978, *Environmental News* reported that the EPA monitored fallout levels in most states had dropped down to the range expected of natural radioactivity. EPA's air monitoring stations in Alabama, Tennessee, and Georgia still report readings indicating traces of fallout. Even these levels are "decreasing rapidly."

But on March 30, 1978, *Environmental News* reported traces of fallout radioactivity in milk! Tests on milk from Montgomery, Alabama, and Little Rock, Arkansas,

> showed trace amounts of fallout which may be attributed to the recent Chinese nuclear test. The concentrations of iodine-131 were 36 picocuries per liter of milk at Montgomery and 20 picocuries per liter at Little Rock. Thirty-seven other milk sampling stations in EPA's nationwide, pasteurized milk network showed nondetectable levels of fallout. Department of Energy facilities have also reported traces of iodine-131 in milk samples from Dayton, Ohio; Aiken, South Carolina; and Oak Ridge, Tennessee.

Here is the fountain effect in action, followed by the horseshoe pattern, as shown on the temperature charts. It just won't quit!

Bomb No. 23
December 14, 1978

Here we go again! A nice Christmas present to the United States, from the People's Republic of China!

Environmental News, Friday, December 22, 1978, reported nuclear debris passing over the United States:

> The Agency [EPA] today reported that the latest forecast prepared by the National Oceanic and Atmospheric Administration indicated the radioactive debris from the Chinese nuclear detonation of December 14, arrived over the U.S. Pacific Northwest at altitudes above 10,000 feet on Thursday, December 21. This debris is expected to reach the East Coast by Saturday night or Sunday. Due to the long travel time across the Pacific, the radioactivity is expected to be widely dispersed and at very low concentrations.

> Some detectable increases in radioactivity in ground level air may occur at times during the next two weeks as portions of the radioactive debris cross the United States. Detectable radioactivity may also be brought to the ground in rain or snow during this time.

If you boil your drinking water to get rid of the chlorine or pour it off while it's still hot to also get rid of the fluoride, you have done all that you can do. You must not use aluminum pots or pans for this purpose. Cast iron will do the job perfectly! Any cast iron utensil is fine, as flourine has an affinity for cast iron.

Does your throat tickle? Do you cough a lot? Does your neck hurt down low where your thyroid glands are? Well, that iodine-131 doesn't just land on pastures. Nor do the other radioactive particulates.

What the fission products attributed to the Chinese bombs do to people has not been determined as yet. Their life and half life is also unknown. (See the fallout chart [Table 12], courtesy of the Oregon State Health Division, Radiation Control Section.) The bombs have been popping off for a long time. There are also some whoppers that were never mentioned before anywhere. *They* all contributed their own light or heavy fallout to all parts of the world. Now let's go back to what EPA had to say about bomb No. 23.

> EPA's air and precipitation monitoring stations, on alert status in each State since last Friday, December 22, 1978, have not reported any radioactivity attributed to fallout. The Agency's nationwide surveillance system will continue monitoring radiation levels very closely, although EPA does not anticipate any cause for concern during passage of radioactive debris over the United States. No significant exposures are expected from direct radiation, breathing of contaminated air, or exposure to rainfall.
>
> EPA has made arrangements to begin next week collecting special milk samples nationwide from a pasteurized milk network, operated in conjunction with the Food and Drug Administration and State health agencies. Experience has shown that radioactive debris brought down to pastures by rain, with resulting contamination of milk, is the most important way by which fallout reaches man. The potential for such contamination at this time of year is lessened, since most dairy herds around the country are on stored feed.
>
> The Federal Aviation Administration has informed EPA that no hazard is expected to air commerce due to air space contamination.

BOMBS OF 1979

On Tuesday night, February 6, 1979, a combat crew of *ours,* from North Dakota, launched a missile down the Pacific missile range. On February 12, 1979, winds of 96 mph swept four men off the deck of a ship, 400 miles off Coos Bay, Oregon. Planes were wrecked at the Astoria Airport, the Hood Canal Bridge in Seattle was blown apart and sunk. Seattle had a cloudburst of rain. Hawaii, on February 22, 1979, according to the National Weather Service, established records for one-hour, twenty-four-hour, and forty-eight-hour periods of rainfall, with 3.04 inches in one hour; 22.31 inches in twenty-four hours, and 28.65 inches in forty-eight hours! Damage to roads and highways on the Island of Hawaii amounted to at least $1.5 million.

In May, we found that 488 previous underground nuclear tests had been conducted at the test site about ninety miles north of Las Vegas, Nevada. (These in addition to the aboveground tests as related in "Nevada Fallout"!) Huge caverns have been blasted deep in the earth. The last one was a dud, but at 2,000 feet down we were assured that it presented no threat to the surface; the detonating circuits have all been disconnected. But aren't 488 huge caverns down there apt to hollow out things enough to cause an earthquake, or just a collapse? There must be a huge vacuum chamber down there. What a hole in the ground that will be! Then, of course, radioactive particles will poof out, like 488 bombs going off all at once! A more virulent bomb could not be devised.

God and the scientists only know what happened on Bikini Island from 1946 through 1958, and the scientists aren't talking. Twenty-three weapon tests were made during that time, including a 15,000 MGT hydrogen (fusion) bomb! How many of those tests were atmospheric weather busters? No one says. As the world rotates west to east, what nation or hemisphere is going to be under the cascading, radioactive waterfall effect when it comes plunging down, perhaps weeks, months, or years later, depending on how far out it punched the atmosphere and at what altitude it or they were detonated? Man, that can be cold with violent winds! Like Sky Lab, the lower it is pulled by gravity, the faster it comes down. Every two weeks around the world—what a crop dusting! We hope and pray that the outer Van Allen radiation screen was not ruptured. That big one was sure to do it. Excessive dosage of cosmic rays plunging through a fifty- or hundred-mile opening of the outer belt could cause worldwide drought; a small hole, a localized one.

FRANCE ALSO BLEW IT

Kay Treakle reports in the *Green Peace Examiner,* and used here with permission of David Rinehart, editor, that French President François Mitterand announced shortly after his election that France would suspend underground nuclear tests at Moruroa, an atoll halfway between Australia and Chile in French Polynesia, until a thorough review of the program could be conducted.

> The people of the South Pacific breathed a sigh of relief. France has tested nuclear weapons above and below Moruroa since 1966. In the past two years, the French have been developing neutron bombs as well as warheads for their version of long-range multiple-targeted missiles.
>
> Barely five days after the announcement was made hardly the time span for a serious review—Mitterand declared France would continue the tests as scheduled. In making this decision, France has again ignored strong protests from South Pacific nations, including New Zealand and Australia. France has also shown disregard for the direct victims of the tests, the Polynesian people, whose environment has been contaminated from radioactive fallout, and whose islands have been turned into near police states by French colonialism and heavy military presence.
>
> In 1960 France carried out its first atmospheric test in the Sahara. Subsequently three more atmospheric tests and thirteen underground tests were completed before two events precipitated the move to the South Pacific. Algeria's hard-won independence from France in 1962 and a miscalculation of weather conditions that sent radioactive fallout wafting across the Mediterranean to Europe forced the French to renege on their promise made in 1961 that "no nuclear tests will ever be made by France in the Pacific Ocean."
>
> ### *The Move to Moruroa*
>
> In 1962 the atolls of Moruroa and Fangataufa were chosen as test sites "for their distance from highly populated areas," according to the French government, since there were fewer than 40,000 people around Moruroa. The Centre d' Experimentation du Pacifique was established to carry out the tests. Despite the protests of the Polynesian people, the CEP brought in thousands of French Legionnaires to construct air fields, wharves, and military support facilities.

While the governor of Polynesia promised that not a single particle of radioactive fallout would ever reach an inhabited island, Albert Schweitzer warned in a letter to the leader of the Polynesian Territorial Assembly that "those who claim these tests are harmless are liars!" Events would prove the latter correct.

By 1966 the CEP was ready to announce its first test and indicated that a danger zone would be established around Moruroa and Fangataufa. When it was pointed out that seven inhabited islands were within the zone, the French didn't move the tests, but merely reduced the area of the danger zone.

The danger zone proved meaningless anyway when the CEP broke its promise that "no bombs would be detonated unless the winds were blowing toward the southern portion of the ocean, where there are no islands."

On September 9, 1966, President de Gaulle visited Moruroa to observe a scheduled atmospheric nuclear test. The explosion was postponed, though, since both sea-level and high altitude winds were blowing toward the islands of the Societies, the Tuamotus, the Cooks, the Samoas, and Fiji. De Gaulle ordered the test to be held the following day in spite of the wind direction, as he was a busy man. To no one's surprise, the blast sent high levels of radioactive fallout as far as Western Somoa, over 2,000 miles to the west.

The all too familiar mushroom clouds continued to rise over Polynesia. There were forty-one atmospheric tests from 1966 to 1974, most of them at Moruroa, since Fangataufa had become too contaminated to use!

International opposition to the tests continued to rise too; but as it did, the French stopped releasing public health statistics for Polynesia. The governments of New Zealand, Australia, and Fiji tried in vain to obtain information regarding the levels of radioactivity caused by the tests, but were told only that "the tests are harmless." In 1971 Peru threatened to break off diplomatic relations with France because of the Pacific tests, and did so in 1973. In 1972, the governments of New Zealand and Australia petitioned the International Court of Justice in the Hague for an interim injunction against the tests. The Court censured France and urged that the tests be stopped after hearing evidence that the radioactive debris was being deposited on Australia and New Zealand. France responded that the Court had no jurisdiction over French national security.

Finally, international pressure, bolstered by the direct action protests of Green Peace and other dedicated environmentalists, forced the French to move their tests underground. For the next four years, the underground tests were conducted with only a minimum of active opposition. But two accidents in 1979 once again focused attention on Moruroa. Two work-

> ers died on June 6, when an explosion ripped through an underground laboratory; their bodies were so irradiated that they had to be shipped back to France in lead caskets. The blast sent plutonium across the atoll.
>
> Later that month a bomb jammed halfway down the 2,600 foot shaft into which it was being lowered. Engineers detonated it anyway. Immediately after the blast, a mile long fissure appeared on Moruroa's surface. Three hours later, a section of the outer wall of the atoll collapsed into deeper water, causing a tidal wave that swept back and injured seven people.

Nuclear tests such as these have so weakened Moruroa that substantial amounts of radiation are leaking directly into the sea. Now the French are considering doing future testing in the lagoon. Unless a halt is called to this nuclear madness, France will continue to bomb the Polynesian Islands, oblivious to the health and environmental as well as social damage it is doing to the people of the Pacific and the world and its oceans. It's very obvious that *they* couldn't care less!

4

Civilizations' Rush to Suicide

WILL OURS BE THE TENTH CIVILIZATION TO FALL?

Now for a bit of philosophy and history before going into the effects of nuclear weapons. Civilizations before the twentieth century had no effect on the weather, but they effectively eliminated themselves! Will ours be the next to fall?

It is just as well that the great men of the past could not foresee the future. Abraham Lincoln, our great President, once said: "Destruction of our country cannot come from abroad, but . . . if destruction be our lot, we must ourselves be its author and finisher. As a nation of free men, we must live through all time, or die by suicide."

We seem to be well on the way, and, of course, we all know that destruction of our country can come from abroad, via the intercontinental ballistic missile.

As for the suicide part, the excerpt below, reprinted with permission of the author, Frank Flick, president of the Flick-Ready Corporation, is of utmost importance to us all in its attempt to show that the United States need not be the tenth civilization to fall.

> "A state chiefly prospers and flourishes by *morality* and *well-regulated family life*, by respect for religion and justice, by moderation and equal distribution of public burdens, by the progress of the arts and of trade, by the abundant yield of the land—by everything which makes the citizens better and happier."
>
> —Rerum Novarum, Pope Leo XIII
>
> These words, though written eighty-seven years ago, still retain their powerful force and deep wisdom. I italicize the

words because, to me, they pinpoint the basic cause for the troubles our civilization faces today and will continue to face. . . .

Nine great civilizations before ours rose, flourished, decayed, and died. Why? Each of the nine reached its peak in about 200 years. Each one seemed invincible, yet crumbled in the face of competing civilizations.

Fallen Civilization No. 1: Babylon

Around 4,000 years before Christ, old Babylon flourished in Asia. Its capital city was said to have been built by Nimrod, grandson of Noah. Renowned for astrology, astronomy, the duodecimal number system, measures of length and weight, the sun dial, and an early calendar, Babylon was brought down by oppressive taxes, political in-fighting, and *moral* decay!

Fallen Civilization No. 2: Egypt

From the fertile Nile delta, Egypt dominated the known world. The Pharaohs built lasting monuments such as the Great Giza pyramid 3,000 years before Christ. Cultural achievements included astronomy, mathematics, medicine, and the arts. But corrupt factions fought each other instead of northern barbarians and again *moral* decay brought disaster.

Fallen Civilization No. 3: Assyria

[Assyria] ruled all of Western Asia with merchants and traders dominant everywhere. Built the world's first roads, constructed aqueducts and irrigation canals. Used cotton for garments, invented the postal system and coinage. But government became too large and crushed the people with *taxes*. Wars sapped economic strength and internal revolts began. Finally, Indo-Europeans looted all, including the glittering capital, Ninevah. *Moral* decay had weakened all defenses.

Fallen Civilization No. 4: Egypt

In the era just preceding the birth of Christ, Egypt again rose and conquered many nations. Art, architecture, literature, and science flourished until internal dissension and oppressive *taxes* caused by *moral decay* took their inevitable toll about 1150 B.C.

New Babylon, No. 5; Phoenicia, No. 6, and the Persian Empire, No. 7, rose and fell victims to *moral* decay in the same way.

Fallen Civilization No. 8: Greece

The ancient Greek city-states of Athens and Sparta emerged as the first models of democratic and of totalitarian governments. In Athens, the first constitution was written about 600 B.C. All free men could vote and hold office. In Sparta, children belonged to the state, as did almost everything else.

The city-states turned into imperialistic bureauracies and

taxed the people heavily. Upon the death of Alexander, internal warfare, sparked by *moral* decay, made the rise of Rome inevitable.

Fallen Civilization No. 9: Rome

A worldwide civilization rose to provide *pax romana* for centuries. But Rome began to weaken when politicians resorted to bribery and subsidies to win power, and decadent pleasures replaced individual industry. Handouts of bread and circuses for entertainment of the masses became the order of the day. Finally, barbarians could be held off no longer and *morally* decayed Rome was sacked, leading to 1,000 years of the Dark Ages. The details may differ but the underlying reason for the fall of all nine civilizations is clear. All of them suffered from *moral* decay, which led to political, economic, and social decay.

I know of only one antidote for moral decay, individual moral responsibility—taken early enough and in good sized doses.

When the sense of individual moral responsibility disappears, leaders succumb to a tyranny of the majority. To please the people and to win reelection, politicians do what is popular instead of what is right. They propose easy, short-range solutions to problems requiring statesmanlike vision and personal courage. They follow public opinion polls rather than the Constitution and their consciences.

The people find it easier to accept *handouts,* than to act with self-reliance and soon begin to fight among themselves to get the largest share.

The people increasingly turn over more and more power and personal freedom to the government in return for subsidies. As government bureaucracies grow, everyone increasingly comes under their control.

Consider the words of Alexander Fraser Tytler, a noted Scottish historian, some 200 years ago. "A *democracy cannot exist as a permanent form of government. It can only exist until the voters discover that they can vote themselves largess from the public treasury. From that moment on, the majority always votes for the candidate promising the most benefits from the public treasury, with the result that a democracy collapses over loose fiscal policy—always followed by a dictatorship.*"

James Madison in his essay No. 10, *Federalist Papers,* spoke out for a republic over a democracy, saying, "A pure democracy . . . can admit of no cure for the mischiefs of faction. A common passion, or interest, will, in almost every case, be felt by a majority of the whole; a communication and concert result from the form of government itself; and there is nothing to check the inducements to sacrifice the weaker party or an obnoxious individual. Hence, it is that such democracies have ever been found incompatible with personal security or the *rights of property;* and have in general been as short in their lives as they have been violent in their deaths."

Individual incentive, creativity, and activation decline as influence becomes more important than excellence as "something-for-nothing" replaces individual self-reliance.

The basic problem of the rest of this century, I believe, is how to regenerate the sense of individual *moral* responsibility in our private and public lives. It is clear that the way to improve society is to improve the individuals who make up that society and vice versa!

To me, this means turning once again to the only source of knowledge, meaning, and strength that endures—our faith in God and the Judeo-Christian tradition.

The principle of individual moral responsibility is the well-spring of honesty, integrity, self-respect, true compassion, and charity which results in less suffering and greater happiness for the greatest number.

Reversing the trend toward moral decay will end our drift toward economic and political decay, because the vast majority of men and women will exercise moral responsibility.

In the final analysis, the fate of nations and empires is determined by how each individual lives his or her life.

I invite businessmen, leaders, and all concerned with the survival of our civilization to join in a "new crusade," to restore the moral health of our people, our nation, and the world.

As Alexis de Tocqueville said about 150 years ago: *"America is great because she is good, and if America ceases to be good, America will cease to be great."*

The Vengeance of Yahweh

Woe to the legislators of infamous laws,
to those who issue tyrannical decrees,
who refuse justice to the unfortunate and cheat the
poor among the people of their rights,
who make widows their prey and rob the orphan.
What will you do on the day of punishment,
when from far off, destruction comes?
To whom will you run for help?
Where will you leave your riches?
Nothing for it but to crouch with the captives and to fall
with the slain!
Yet his anger is not spent,
Still his hand is raised to strike!

Isaiah 10:1-4

History need not repeat itself. It's exciting and *we can profit* from reading it! The romance and disasters of the past, the rise and fall of great civilizations, as presented here, need not be repeated. The warning is plain! It gives us a good look at how *they* are blowing our political weather as well as the air above us. It shows the cause and cure. Is America on the way to suicide? Was World War II Armageddon? *Is it later than we think?*

Now destruction of our country *can* come from abroad. Our government, in its quest for power or peace, and our so-called representatives and those of other nations on earth are well on the way to being *the authors and the finishers* of not only the United States but of themselves and the entire world! Developing atomic weapons has got to stop! No one dares to use the ones now in existence. President Eisenhower, in his farewell address, when he left office, said, "Beware of a military-industrial complex"!

Our scientists and the military-industrial complex are dashing pell-mell toward *suicide!* No place on earth is beyond the reach of ICBMs. What's worse, the whole world is *now,* by atmospheric diffusion, blanketed by radioactive debris clouds. A complete shell around the earth! It's sixty miles thick!

Mankind in this century has made more scientific, medical, and industrial progress than in the last ten thousand years! *Nothing* must be allowed to impede this development. Now just what have we done with the atom?

Pigs are being raised that mature to the approximate size and weight of a person. Plutonium-239 is administered under controlled laboratory conditions to determine just what a safe dose of radiation is for people! There are radioisotopes and Laser beams that with electron microscopes allow surgeons to perform surgery that they wouldn't have dreamed of trying before. This was developed from death-ray research! Medicine uses nuclear power. Breeder reactors produce more fuel than they consume for home heating, fuel for industry. Who needs oil, coal, or other nonreplaceable resources? It's how you use it, not what it is; risks can be eliminated.

Let's learn how to use nuclear power safely. Let's learn how to utilize the nuclear waste and surplus breeder reactor fuel. A brick of waste that still produces intense heat can be used as fuel for a Stanley Steamer car, with the gas tank full of water instead of gasoline or diesel fuel! The Stanley Steamer was a success but was killed in its infancy by the oil companies. Incidentally, if fluoride will make water burn with a nice, clean, blue flame, like alcohol, what's holding back our researchers? Engines are being developed in Brazil that will run on pure alcohol! Of course, a radioactive brick used as fuel for a steam car might last a thousand years! A miniature nuclear power plant like that, however, would be a no-no, as far as oil, coal, and electric advocates are concerned! *They* want to close down all recycling and research labs. Something useful might be found! That's the first thing President Carter did in April of 1977. Did the oil cartel elect the man? He also ordered the closing down of all hydroelectric projects, even those nearly completed. What better way to close down the store? I'll bet Teddy Roosevelt turned over in his grave when they gave away the Panama Canal! All other nations that have the capability are busy, busy, researching, recycling nuclear wastes. *We can* produce our way out of any shortage and solve any problem if given free rein! The real problem is bureaucratic rules with the force of *law!* That's *unconstitutional!* Let's dispel whatever gloom we have created so far by lampooning bureaucracy.

"THE ENVIRONMENTAL IMPACT OF CREATION"

Andrew J. Hinshaw, Republican congressman from California, is the author of this commentary reprinted below from the *Congressional Record.*

> In the beginning God created heaven and earth.
>
> He was then faced with a class action lawsuit for failing to file an environmental impact statement with HEPA (Heavenly Environmental Protection Agency), an angelically staffed agency dedicated to keeping the universe pollution free.
>
> God was granted a temporary permit for the heavenly portion of the project, but was issued a cease and desist order on the earthly part, pending further investigation by HEPA.
>
> Upon completion of the construction permit application and environmental impact statement, God appeared before the HEPA Council to answer questions.
>
> When asked why he began these projects in the first place, he simply replied that he liked to be creative.
>
> This was not considered adequate reasoning and he would be required to substantiate this further.
>
> HEPA was unable to see any practical use for earth, since "the Earth was void and empty and darkness was upon the face of the deep."

Then God said, "Let there be light."

He should never have brought up this point since one member of the council was active in the Sierra Angel Club and immediately protested, asking, "How is light to be made? Will there be strip mining? What about thermal pollution? Air pollution?" God explained the light would come from a huge ball of fire.

Nobody on the council really understood this, but it was provisionally accepted assuming (1) there would be no smog or smoke resulting from the ball of fire, (2) a separate burning permit would be required, and (3) since continuous light would be a waste of energy, it should be dark at least one half of the time.

So God agreed to divide light and darkness, and he would call the light Day and the darkness Night. (The council expressed no interest with in-house semantics.)

When asked how the earth would be covered, God said, "Let there be firmament made amidst the waters, and let it divide the waters from the waters."

One ecologically radical council member accused him of doubletalk, but the council tabled action, since God would be required first to file for a permit from the ABLM (Angelic Bureau of Land Management) and further would be required to obtain water permits from appropriate agencies involved.

The council asked if there would be only water and firmament, and God said, "Let the earth bring forth the green herb and such as may seed and the fruit tree yielded fruit after its kind, which may have been itself upon the earth."

The council agreed as long as native seed would be used. About future development God also said, "Let the waters bring forth the creeping creatures having life, and the fowl that may fly over the earth."

Here again, the council took no formal action since this would require approval of the Game and Fish Commission coordinated with the Heavenly Wildlife Federation and Audubongelic Society.

It appeared everything was in order until God stated he wanted to complete the project in six days.

At this time he was advised by the council that his timing was completely out of the question. . . . HEPA would require a minimum of 180 days to review the application and environmental impact statement, then there would be the public hearings. It would take ten to twelve months before a permit could be granted.

God said, *"To hell with it!"*

Now back to serious stuff.

Benjamin Franklin said: "Before too long, men will learn how to levitate huge masses of matter." Isn't it possible the

Egyptians and others before them knew how? How about the magnetic trains of Japan? What's next? Barrier after barrier falls, or is it just rediscovery? Science is riding the rocket. Nikola Tesla really put the twentieth century on the track. He is honored by being in the Scientific Hall of Fame. We go step after step up the ladder of achievement. However, testing of atomic weapons in the *atmosphere* and under the *ocean has got to stop!* In fact testing *atomic weapons* had better stop altogether!

The jet streams of the troposphere were blasted from their formerly dependable courses and must now be hunted down. The U.S. Weather Service performs this service for the high altitude aircraft. The intercontinental jets were and are able to cut their fuel burn in half, if they can find a jet stream going their way! Manmade metals, never known on earth before, are now serving mankind. Some serve as shields to radiation. They will not absorb it! Titanium and zirconium are two new ones with many uses. Miniturization is also astounding! Many developments are side effects of atomic research and moon shots.

"Those people who only insult anything they do not understand, are not reasoning beings, but simply animals, born to be caught and killed, and they will quite certainly destroy themselves by their own works of destruction" (2 Peter 2:11-12).

Now just suppose someone should decide to initiate a first strike or, as did happen one morning, a computer foul-up showed North America to be under missile attack by a foreign power. Evidently the *hot line* prevented a counterattack. Computers *could be the death of humanity!*

5

They Blew Our Weather—Now They Are Killing Our Oceans

WHY ARE WHALES BEACHING?

On June 16, 1979, a pod of forty-one sperm whales beached themselves on the Oregon coast. Kindly fishermen and tugboat captains towed some back out to sea, only to suffer the disappointment of seeing them swim back and struggle to get out of the water. (See Plates 5-9.)

Now science is unable to feel or measure pain in people or animals, but those whales must have been hurting pretty bad to come ashore and try to walk like people. That water somewhere along the line must have been pretty *hot!* Marine biologists of Oregon State University were unable to find any illness or disease in any of them. When I asked about the possibility of radiation poisoning, I got the old bureaucratic runaround. I was told that research would be completed by the first of the year and I would be the first to know at that time. Well, the first of the year rolled around and, of course, the paper was not completed, thorough research does take time. On a query, I was told it would be finished soon and published in a marine science journal; that I was on their mailing list and would receive a copy. It is now June, 1980, and nothing. I am positive that it was radiation poisoning or I would have had a quick denial, and they are such nice people! A clincher: I was told recently by a coastal fisherman that some of his buddies got themselves several nice whale roasts. He said, "The FBI was sent after us, and when we asked what was wrong with the meat, we were told that it was contaminated!" That should prove my theory, if we never hear from the marine biologists.

On June 22, 1980, the *Morning Oregonian,* of Portland, printed the following article under the byline of Nicholas D. Kristof, an *Oregonian* staffer.

WHALE BEACHING DEFIES ANSWER
But Death Cause Determined

The sun was sinking over the Pacific Ocean a year ago last Monday when 41 enormous sperm whales came swimming toward a deserted beach near Florence on the central Oregon coast.

As the tide went out, the whales were left trapped by a sandbar, mired in the wet sand. All the whales eventually died, although some lingered for two days after the stranding. Thousands of tourists descended upon the isolated beach to watch as the giant mammals stared out from pools of their own blood.

What was tragedy for the whales and for those who saw them die was also a tremendous windfall for scientists, who never before had access to such a large sperm whale stranding.

Tissue samples were taken from all the whales, and skulls and other body parts from many of them, and scientists agree that the whales in effect cooked to death in their own body heat.

Deborah Duffield, an assistant professor of biology at Portland State University who has studied the whale tissues extensively, explained that the whales produce an enormous amount of body heat because of their metabolism.

Normally this body heat is carried away by the water, particularly from the flukes and flippers where there is less blubber, but out of the water they quickly overheat because air is a much slower conductor of heat than water.

Scientists now know that the most important thing to do when called to a stranding is to keep the flukes and flippers wet, said Tag Gornell, director of the Marine Animal Resources Center in Seattle. It may not help save the whales —because no true whale has survived a beaching—but it will preserve tissues so they can be studied later, Gornell said.

There is much less agreement among scientists as to the reason for the stranding. Whales have remarkable accurate sonar systems and they normally stay in deep water. Possible explanations for the stranding include the following:

—Squid are a principal food of sperm whales, and area fishermen reported that squid had moved much closer to shore at the time of the stranding. Denise Herzing, a research assistant at the Oregon State University Marine Science Center in Newport who spent last summer examining stomachs from the stranded whales, said some of the whales had eaten squid shortly before stranding on the beach. While eating the squid, the whales might have become frenzied and lost their bearings.

—The whales might have been following a dominant male who was diseased or suffering from parasites. No strong evidence of disease or parasites was found, but many organs and tissues had been *cooked by the heat* by the time scientists analyzed them [emphasis added]. Also, the largest male found was only 28 years old, 38 feet long and 25 tons, while a dominant male for a herd probably would be older and larger, Duffield said.

—The whales were searching for a dominant male who had already died. A 60-foot bull sperm whale had washed up at Winchester Bay six weeks earlier, but he had died at sea and it is unclear why the rest of the herd would enter shallow water at Florence to look for him.

—An earthquake had interfered with the whales' sonar systems. An earthquake was recorded in the area earlier the day the whales stranded, but scientists point out that there have been other earthquakes with no strandings and other strandings with no earthquakes.

Even scientists grasp at straws when no believable answer is evident. According to scientists and researchers, and there were many, the whales beached themselves without a good reason. Make your own deductions from what you read here! As I suspected, the whales *were* cooked to death. (But in their own body heat?) Now bleeding to death because their body structure could not support their weight would have made a more plausible story. The rupture of blood vessels, arteries, and organs accustomed to being supported by the ocean, just gave up.

No one has denied my declaration that they were victims of an underwater detonation of an atomic or fusion bomb. The entire production of radioactive debris can be carried thousands of miles. Due to the density of the water, it is concentrated and very deadly to *anything* that swims through it or lives on the bottom as it passes over. Death resulting from an underwater detonation may be very painful and lingering! The debris can be carried into major ocean currents and then follow the Japanese floats onto the beaches and into estuaries all along the coast lines.

That pod of whales was a hot potato, a no-no. The research was done secretly; too many protesters now! Wait until Green Peace gets hold of this! The whales may have been in the vicinity of a bomb test, or were swimming through an ocean current carrying the entire load of debris from an underwater detonation. Depending on the depth of burst (DOB), it could do enormous damage to all sea life. Was the haste to burn the carcasses more than just for sanitary reasons? Several byproducts firms offered to remove the carcasses free of charge. But no, tax money had to be expended to cover up, and I don't mean just to bury them!

A portable cannery could have been set up close by and the meat processed for the world's poor people, or our own hungry people on welfare. Whale meat tastes like the finest steak you ever tried. There never was a tough whale! If not for human consumption, pet food is in demand in all groceries everywhere; the whale meat might have saved a lot of horses from going to the dogs except that they too have been feeding on irradiated pastures!

Scientists, conservationists, and government officials assembled in Washington to work out solutions to the complex problems related to the disappearance of the once numerous sea turtle. Participants developed an international action plan for protecting the species—the first time such a project has been undertaken.

One of the worst examples of the commercial depletion of the sea turtle is found in Mexico, where the once abundant East Pacific green turtle is in danger of disappearing. Scientists did not mention that part of the reason for the scarcity might be radioactive currents along the coast of Mexico! Turtle meat and soup are also a gourmet's delight. (See Fig. 25.)

Now these weren't the only cases of this sort! Our marine biologists and those of Canada are very sharp about these things and I feel sure that they are well aware by now of exactly what the problem is. On July 17, 1979, I wrote to the superintendent of the Bureau of Fisheries, Viking Building, St. Johns, Newfoundland, Canada ALC 5T7, to inquire about the fate of the 170 pot head whales that beached themselves and died within a matter of hours. Fisheries personnel and many fishermen in boats herded another 60 whales back into deep water. But more whales swam ashore. There was no answer to two letters and I do not expect any! Those 170 whales that beached themselves near Point Au Gual, on Newfoundland's south coast, weren't just in fear of something; they were being *cooked alive!*

The whales, ranging in size from six to about twenty-five feet, came ashore in two pods, one of about 150 animals, and another of 25 or 30. Officials at that time stated that the cause of such suicidal action was unknown!

FIGURE 25

The Center for Environmental Education launched a program in 1979 to protect the dwindling sea turtle population.

UNDERWATER DETONATIONS

Underwater detonations of atomic weapons will, if they haven't already done so, poison or kill enormous amounts of sea life of all kinds! Operation Baker (shallow underwater) created a 94-foot high wave at 1,000 feet from ground zero! It traveled outward at high speed, the depth of inundation being twice as high as the wave height. This BAKER test at Bikini in July, 1946, of 20 KT. was only 200 feet deep, and called shallow! This was only the beginning of thermal and nuclear underwater bursts! (See Fig. 26.)

Note the ships in the shock wave area. There might have been some in the center at ground zero. If so, no mention was made as to how they came out. That would be interesting to know, but no doubt it would be a closely guarded secret. In Fig. 27 one of the ships looks as if it is climbing up the stem of the mushroom column. The Air Service photographers were really sticking their necks out to take the pictures. No one knew what might happen.

FIGURE 26

Condensation cloud after a shallow underwater explosion. *(From* ENW)

FIGURE 27

Formation of the hollow column in a shallow underwater explosion; the top is surrounded by a late stage of the condensation cloud. (*From* ENW)

The Effects of Nuclear Weapons (ENW), the government study published in 1977, reports the following concerning underwater explosions:

> If the depth of burst is not too great, the bubble remains essentially intact until it rises to the surface of the water. At this point the steam, fission gases, and debris are expelled into the atmosphere. Part of the shock wave passes through the surface into the air, and because of the high humidity the conditions are suitable for the formation of a condensation cloud [Fig. 26]. As the pressure of the bubble is released, water rushes into the cavity, and the resultant complex phenomena causes the water to be thrown up as a hollow cylinder or chimney of spray called the "column" or "plume." The radioactive contents of the bubble are vented through this hollow column and may form a cauliflower-shaped cloud at the top [Fig. 27].
>
> In the shallow underwater (BAKER) burst at Bikini, the spray dome began to form at about 4 milliseconds after the explosion. Its initial rate of rise was roughly 2,500 feet per second, but this was rapidly diminished by air resistance and gravity. A few milliseconds later, the hot gas bubble reached the surface of the lagoon and the column began to form, quickly overtaking the spray dome. The maximum height attained by the hollow column, through which the gases vented, could not be estimated exactly because the upper part was surrounded by the radioactive cloud [Fig. 28]. The column was probably some 6,000 feet high and the maximum diameter was about 2,000 feet. The walls were about 300 feet thick, and approximately a million tons of water were raised in the column.
>
> The cauliflower-shaped cloud, which concealed part of the upper portion of the column, contained some of the fission products and other weapon residues, as well as a large quantity of water in small droplet form. In addition, there is evidence that material sucked up from the bottom of the lagoon was also present, for a calcareous (or chalky) sediment, which must have dropped from this cloud, was found on the decks of ships some distance from the burst. The cloud was roughly 6,000 feet across and ultimately rose to a height of nearly 10,000 feet before being dispersed. This is considerably less than the height attained by the radioactive cloud in an air burst. [No thought was given as to what was happening to the ocean bottom.]
>
> The disturbance created by the underwater burst caused a series of waves to move outward from the center of the explosion across the surface of Bikini lagoon. At 11 seconds after the detonation, the first wave had a maximum height of 94 feet and was about 1,000 feet from surface zero. This

moved outward at high speed and was followed by a series of other waves. At 22,000 feet from surface zero, the ninth wave in the series was the highest with a height of 6 feet.

It has been observed that certain underwater and water surface bursts have caused unexpectedly serious flooding of nearby beach areas, the depth of inundation being sometimes twice as high as the approaching water wave.

FIGURE 28

The radioactive cloud and first stages of the base surge, following a shallow underwater burst. Water is beginning to fall back from the column into the lagoon. (*From* ENW)

I wonder if many islanders on the nearby islands lost their homes or their lives? The debris can be carried thousands of miles by any of the various strong underwater currents!

The English have dumped 5,000 steel barrels full of radioactive waste in the mid-Atlantic, and a steel barrel has a very short life in sea water. Other nations, as well as our own, dump radioactive waste in the same location and some in more convenient areas, closer to civilization.

The Green Peace people have risked their lives trying to prevent dumping. Yet, scientists are still trying to bury this waste, instead of trying to find a use for it, or finding a way to neutralize it. It's just possible that radioactive particles may have an affinity for *salt!* Radiation travels with the speed of light!

In the area of Hemlock, Michigan, people have gotten mysteriously ill, animals drop dead, and lakes are covered with a fluorescence of some sort that retreats when stirred with the hand. Some wild mutations have occurred among poultry.

Dr. John L. Isbister, Disease Control officer for the state of Michigan, states that an ongoing in-depth study will be concluded shortly. The Dow Chemical Company of Midland, Michigan, is cooperating to determine the possibility of ground water contamination. Fluorescein ($C_{20}H_{12}O_5$), a yellowish red crystalline compound, used in medicine and dye stuffs, may be suspect. Or it may be fallout from Chinese bombs!

THE ATOM—ENIGMA OR SIREN?

Has science whelped a monster? Or has science just cut the chains that bound the devil? *We can and we must control* that which we have loosed on the world! Will it be our servant, our master, or our executioner?

It certainly is *our enigma!* Remember the old Greek legend of the siren on the rocks, beautiful nymphs that lured sailors to their deaths? Maybe it's that also! Perhaps it's just a phenomenon?

You know what a phenomenon is: If you are out for a ride in the country and see a bull in a pasture and a bull thistle close by with a canary on it, singing, that, my good friends, is perfectly normal. Right? Now if you pass that same pasture again on your way home and you see the bull sitting on the bull thistle, singing like a canary, that, my friends, is a phenomenon! Sorry folks, I don't have a picture of that! Now for those of you who expect to read of great discoveries here, *you will, you will.* I will even read between the lines of declassified government statistics for you, minus the jargon.

You say, what's with this guy? Well, let's assume I'm a nuclear physicist and that I'm going to present to you some of what you want to know about fission and fusion bombs.

The TRINITY test of July 16, 1945, near the town of Alamogordo, New Mexico, had a yield of about 19 kilotons, the equivalent of 19,000 tons of TNT. At that time scientists knew nothing about the results or side effects of nuclear fission or fusion, which information came much later. They noted only the destructive and immediate effects of radiation. What I tell you now should scare the hell out of you. Now *they* know! Radiation at the speed of light, 186,300 miles per second, can be very deadly! A 1 megaton bomb explosion creates blast winds of more than 70 mph as far away as eight miles from ground zero (the point of detonation). Infrared rays are instantaneous and cause most of the burns as far away as twelve miles from ground zero. Ultraviolet rays are the most damaging to the eyes. If you can drop behind a solid wall or curb at the instant of the flash it will mean the difference between life or death! The blast wave will strike soon after, though much slower but just as deadly unless you are below ground level.

In the 1946 ABLE test at Bikini, the bomb was detonated just above the surface of the lagoon (see Fig. 29). Notice the naval vessels positioned for test of damage and survival ability. No mention was made of this, and no data are available at this time. (An air detonation is 100,000 feet or less. Anything above that altitude is an atmospheric test.)

The first atmospheric test, called TEAK, occurred on August 1, 1958, at 252,000 feet, or nearly 48 miles up! There were a series of atmospheric tests, code named HARDTACK, in the Pacific, and ARGUS in the South Atlantic. In the HARDTACK series, two high-altitude bursts, with energy yields in the megaton range, were set off in the vicinity of Johnston Island, 700 miles southwest of Hawaii. See Fig. 30, a photo of the TEAK detonation, photographed in Hawaii, 780 miles away!

Then high-altitude tests began popping off all over, during the FISHBOWL series in 1962. The yield from these ran from 1.4 megaton thermonuclear (fusion) devices to 3 submegaton devices. The STAR FISH PRIME device, with a yield of 1.4 was detonated at about 248 miles up, on July 9, 1962 (GCT).

The three submegaton devices, CHECKMATE, BLUEGILL, TRIPLE PRIME, and KINGFISH, were detonated at altitudes of tens of miles on October 20 and 26 and November 1, 1962.

Within a fraction of a second of detonation of TEAK, purple streamers were seen to spread to the north. Less than a second later, an aurora was observed at Apia in Samoa more than 2,000 miles away, though at no time was the fireball in direct view.

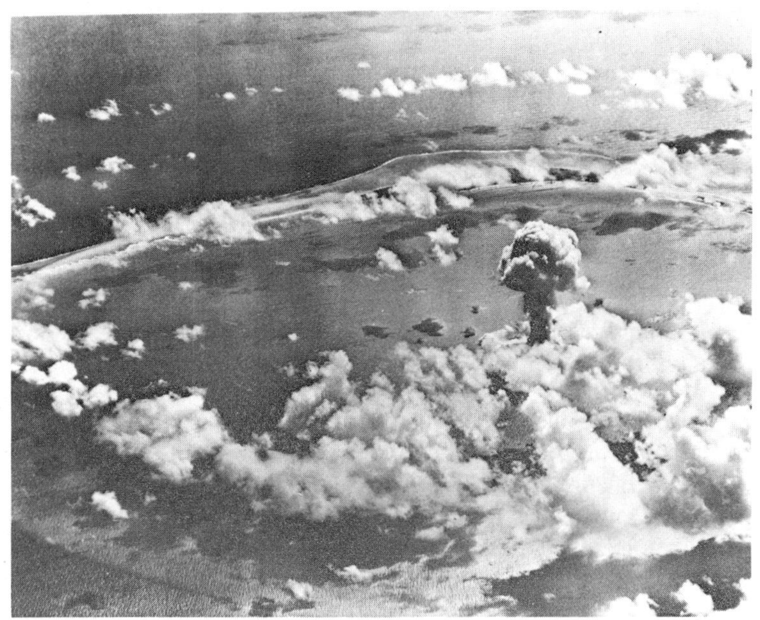

FIGURE 29

The 1946 ABLE test at Bikini. *(U.S. Air Force photo)*

FIGURE 30

Fireball and red luminous spherical wave formed after the TEAK high-altitude detonation. The photo was taken from Hawaii, 780 miles from the explosion. (*From* ENW)

Similar aurora effects occurred after other high-altitude explosions. See photo taken at the start of Oregon's October 12 Columbus Day storm (Plate 3).

The government study *ENW* reports the following:

> At about a minute or so after the detonation, the Teak fireball had risen to a height of over 90 miles, and it was then directly (line-of-sight) visible from Hawaii. The rate of rise of the fireball was estimated to be some 3,300 feet per second and it was expanding horizontally at a rate of about 1,000 feet per second. The large red luminous sphere was observed for a few minutes; at roughly 6 minutes after the explosion it was nearly 600 miles in diameter [See Fig. 30].

Now, what this effect will do or did do is anybody's guess, or, perhaps, various side effects were discerned and not reported, as of small consequence. Were it not for the absorption or reflection of much of the solar ultraviolet radiation by the ozone and the Van Allen radiation belts, life as we know it could not possibly exist except perhaps in the oceans! All atomic fission or fusion radiation could be and probably was reflected from this shell! Then again it is possible that a 15 thousand megaton hydrogen bomb blew open a huge hole in the Van Allen belts, leaving a window for pure, raw solar radiation to shine through as it slowly moves around the world. Could that be the problem in Texas and other worldwide hot spots? Then again it could be the earth moving slowly under the window?

Ozone (O_3), is formed in the upper atmosphere, mainly in the stratosphere, by the action of solar radiation on molecular oxygen (O_2). Small quantities of nitrogen oxides can cause a considerable breakdown of the ozone. Nitrous acid attracted by gravity could cause a great amount of damage to all concrete structures and other building materials, as well as fibers and tissue. Now just how dangerous is this debris and fallout, aside from the metallic junk that is up there?

HOW DANGEROUS ARE THE EFFECTS OF RADIATION?

Nuclear radiation consists of gamma rays, neutrons, beta particles and alpha particles, ultraviolet rays, infrared rays, X rays, and, no doubt, many others that haven't been identified as yet. Some solar Ultraviolet rays on the other side of the Van Allen radiation belts plunge down to earth the second an opening is made by high-altitude detonations! At one mile from ground zero, twenty-four inches of concrete would give no protection from the radiation of a 1 megaton bomb.

A *roentgen* is the measure of quantity of gamma or X rays. A *rad* is the unit of energy absorption applied to all nuclear radiations. It represents the deposition of 100 ergs of radiation energy per gram of absorbing material. Lithium fluoride is used in radiation detectors.

The isotope of uranium or plutonium, exposed to neutrons, creates by fission highly radioactive beta particles and gamma rays. Time of exposure will determine excitability or deadly effect on (animal) tissue. Thermonuclear processes (fusion) produce deuterium and tritium and more high-energy (fast) neutrons than fission explosions. The sea-level density of air normally is 0.9-. Neutrons travel faster than the blast wave and faster than gamma rays. They are emitted *for only an instant,* which results in the rapid melt-down of animal tissue!

The product of neutron interaction with nitrogen is carbon-14, which is radioactive. It emits beta particles of low energy but no gamma rays. Carbon-14 has a long half life (the period of time it takes for it to decay and became harmless to human or animal life)—5,730 years! It emits beta particles very slowly in the form of carbon dioxide. It is readily incorporated in plant life and finds its way into humans and animals. It is estimated that before September, 1961, weapons testing had produced, in addition, 0.65 short tons of carbon-14. The total reservoir of carbon-14 in nature, including oceans and atmosphere, is normally 50 to 80 tons. About half of the weapons producing carbon-14 has dissolved in the oceans. As a result of the *large number* of atmospheric nuclear tests of high-yield hydrogen bombs in 1961 and 1962, the *excess* of carbon-14 in the atmosphere rose to about 1.6 tons. By mid-1969 the excess had fallen to about 0.74 ton. In the course of time, more and more of the carbon-14 will enter the oceans! Poor fish! If there is no great addition of carbon-4 as the result of additional weapons tests—*and, by God, they had better stop*—the level in the atmosphere will continue to decrease. If the rate of decrease of excess carbon-14 observed in the atmosphere between 1963 and 1969 were to continue, the level should fall to less than 1 percent above normal in *40 to 80 years!*

The deuterium isotope, nonradioactive, is produced when neutrons are captured by hydrogen nuclei in water (H_2O), seeming to effectively neutralize the dangerous radioactivity of the neutron. However it's not such a great idea, as the radiation of the neutrons are instantly deadly. Of course, it would make it possible to occupy enemy territory within minutes after a dose by the neutron bomb! The fact that uranium and plutonium are not entirely consumed in a fission explosion leaves some interesting facts to consider.

The uranium and plutonium that escape fission in the nuclear weapon represent a further possible source of residual nuclear radiation. Again, depending on the altitude, time exposure, and whether you dropped to the ground behind sufficient protection at the instant of the flash, you may live to a ripe old age of half life, like some of the radioactive particles, or you may not make it through the day!

However it is not uranium or plutonium that is so deadly. These are the pussycats of the bomb. They supposedly are completely absorbed in an inch or two of air. Of course, there is no air in the fireball. This, together with the fact that the particles cannot penetrate ordinary clothing, indicates that uranium and plutonium deposited on the earth *do not* represent a serious *external* hazard. Even if they actually come in contact with the body, the alpha particles emitted are unable to penetrate unbroken skin. Of course, the blast wave will have tossed you around, and you will no doubt have a lot of broken skin!

It is possible for dangerous amounts of these elements to enter the body through the lungs or the digestive system. For example, plutonium tends to concentrate in the bone and lungs, where the prolonged action can cause serious harm. At one time it was suggested that the explosion of a large number of nuclear weapons might result in such an extensive distribution of plutonium as to represent a worldwide hazard. It is now determined that the fission products of the radioisotope strontium-90, in particular, are a more serious hazard. Of course, all the others will be there worldwide! You cannot hide from radiation! It may pass over a foxhole or a wall, but it acts like water. Radiation fills in behind, if it doesn't pass through obstructions! By the time the less harmful rays reach your area, the deadly ones have been long gone. Of course, they will, like General MacArthur, return—in fallout on the next trip around the world, which takes approximately two weeks, depending on the height of detonation.

High winds from the upper air forced in all directions by the explosions may bring fallout down or scrounge it out of the stratosphere as snow or rain. Snow falls at 200 feet per minute and rain at 800 to 1,200 feet per minute, and this captures far more radioactive particles than a slow fall by gravity. It can be carried in clouds thousands of miles from the bomb burst and deposited on the senders or on lands and people not concerned in any way, except that they did not try hard enough to prevent such a holocaust.

In general, except at ground zero, at any given point away from a surface explosion, some time will elapse between the blast and the fallout! This time depends on the distance from ground zero, the wind velocity and blast wave, and the altitude of the detonation. The first fallout will be light but it increases as the radioactive cloud or clouds lower or intersect thunderstorms on repeated return passes. After, if, and when the fallout is ever completed, the radioactive decay of the fission products will cause the dose rate to decrease—small consolation for those exposed to lord knows how high a dose!

It took five days for the 400-mile wide and 60-mile thick radioactive cloud to reach the West Coast from LocNor testing grounds in the People's Republic of China, there was practically no fallout. The second one took seven days to reach the California coast and was much lower than the first one. That fallout will extend over a long period of time. Months or even years may pass. A radiation dose of 700 rads over a period of ninety-six hours would probably prove fatal in most cases!

WHAT A 15 THOUSAND MEGATON BOMB CAN DO

The effect is no worse than a very severe thunderstorm; that is what meteorologists say about atmospheric bombs. Yet the equivalent of fifteen thousand million tons of TNT was detonated about seven feet above the surface of a coral reef, and the resulting fallout, consisting of radioactive particles ranging from about one-thousandth to one-fiftieth of an inch in diameter, contaminated an elongated area extending over *330 miles downwind* and *20 miles upwind!* It varied in width up to over 60 miles. That's contamination! (See Fig. 31.)

Observations were made of the Marshall Islands area following the high yield of this explosion (BRAVO) at Bikini Atoll on March 1, 1954. According to *ENW,* "a total area of over 7,000 square miles was contaminated to such an extent that

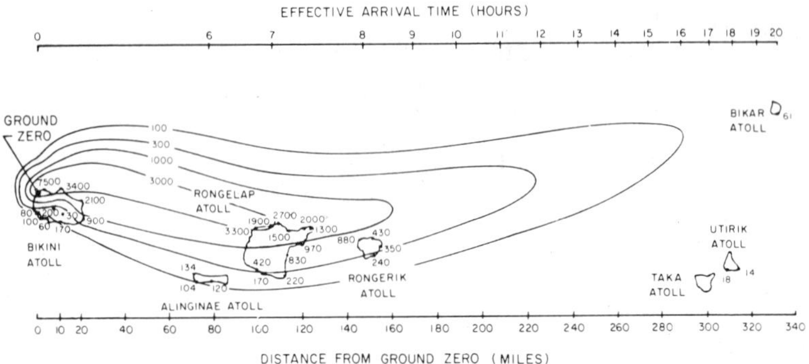

FIGURE 31

Estimated total (accumulated) dose contours in rads at 96 hours after the BRAVO explosion. (*From* ENW)

avoidance of death or radiation injury would have depended upon evacuation of the area or taking protective measures." *Somebody goofed!* Or was it deliberate? Either the effects of high-yield explosions were still unknown or guinea pigs were needed.

THE MARSHALLESE EXPERIENCE

On March 1, 1954, when the high-yield (BRAVO) test explosion took place at Bikini Atoll, the people of the Marshall Islands were not aware of the significance of the fallout. Many of the inhabitants ate contaminated food and drank contaminated water from open containers for periods of up to two days before they were evacuated from the islands (see Fig. 32). What beautiful guinea pigs! (See Fig. 33.) Science can be cruel and indifferent and *deceitful*; it makes me wonder sometimes.

For the Marshallese people, the internal deposition of fission products resulted mainly from ingestion rather than inhalation for, in addition to the reasons given above, the radioactive particles in the air settled out fairly rapidly as powdered coral, or chalklike dust. The belief that ingestion was the chief source of internal contamination was supported by observations on chickens and pigs made soon after the explosion. The gastro-

FIGURE 32

In the Marshall Islands, drinking water is collected from rain that trickles down from the trunks of trees. *(Photo by Willard Price)*

FIGURE 33

What a beautiful guinea pig! *(Photo by Willard Price)*

intestinal tract and the liver were found to be much more contaminated than lung tissue.

From radiochemical analysis of the urine of the Marshallese that were subjected to the early fallout, it was possible to *estimate* the body burdens, i.e., the amounts deposited in the tissues of various isotopes. It was found that iodine-131 made the major contribution to the activity at the beginning, but soon disappeared because of its relatively short radioactive half life of 8 days! Somewhat the same was true for barium-140 with 12.8 days half life, but the activity levels of the strontium isotopes were more persistent. Not only do these isotopes have longer radioactive half lives, but the biological half life of the element is also relatively long.

No elements other than iodine, strontium, barium, and the rare earth group were found to be retained in appreciable amounts in the body. Essentially, all other fission products and weapon residue activities were rapidly eliminated, because of either the short effective half lives of the radionuclides, the sparing solubility of the oxides, or the relatively large size of the fallout particles.

Here we have a great ado about the half life of radioactive elements and also about the biological half life of these elements in bodies, whether they be human or animal. The implication is that with iodine-131 after 8 days it is gone—not to worry. Barium-140, with 12.8 days half life, the same thing. If we use the analogy of a bullet, it's not how long it takes to pass through the body but how much damage it does while it's going through! Also, some fallout radioactive particles are too large to inhale, but what about drinking water? May it not carry them through your body very nicely?

People have sat on or stepped on a needle and had it come out of them years later, far removed from the point of entry, with no harm done. However the thyroid gland has an affinity for iodine-131. Though doctors use it for a scan to determine the gland's condition, they also use it to destroy a portion of the gland if it's overactive! It seems to me that that is inviting cancer of the thyroid? The thyroid controls the endocrine glands, any malfunction of which may result in serious disorders or death! How much to administer so as not to destroy too much? Other elements concerned have an affinity for bone marrow and calcium, and can cause cancer years later!

Neutron rays penetrate anything and last only a fraction of a second! *But* they cause rapid meltdown of *all* animal tissue. The half life of a neutron ray is less than a second, but that's far too long.

Until 1963, no thyroid abnormalities had been detected among the inhabitants of the Marshall Islands that could be attributed to the fallout. In that year, one case was found among the people of the Rongelap Atoll, but by 1966 there were eighteen cases; the total number increased by twenty-two by 1969 and to twenty-eight by 1974, which is twenty years after the explosion! Of the Rongelap people who were exposed, sixty-four (plus one *in utero*) who were on the neighboring Alingina Atoll (see Fig. 31), at the time of the nuclear blast received about sixty-nine rems, or units of biological dose of radiation. The thyroid doses from radioiodines were much larger, especially in children under sixteen years of age. In 1974, there were twenty-two individuals with thyroid lesions among the more highly exposed group and six among the others. There were no definite malignancies in the latter group although there was one doubtful case, which given the time required in a more resistant individual probably would develop as a cancer! All other thyroid abnormalities were benign nodules (and who needs them?), difficulty in swallowing, constant coughing, excessive saliva secretions, toxic goiter, and Graves' disease (bulging of the eyeballs). You are what your glands are!

Most of the lesions occurred in children who were younger than ten years old at the time of the explosion in 1954. Out of nineteen children—the total of those that were on Rongelap—*seventeen developed abnormalities,* including one malignancy and two cases of hypothyroidism. The radiation doses from radioiodine isotopes that had been concentrated in the thyroids of these children were estimated to be from 810 to 1,150 rems. In 1974, a lesion was observed in one of the individuals who had been exposed *in utero;* the thyroid dose was uncertain but it must have been at least 175 rems. The six children in the less highly exposed group who were on Alinginae, received estimated thyroid doses of 275 to 450 rems by 1974; lesions were observed in two cases with one doubtful malignancy. No two-headed monsters or people with four arms or legs were found, or at least they haven't come to light as of 1977.

EFFECTS OF EARLY OR FAST FALLOUT

Life Shortening

The life shortening in a given animal for a specific radiation dose apparently depends on such factors as genetic constitution and on the age and physical condition at the time of the exposure! In other words, guys and gals, tread the straight and narrow.

Retarded Development of Children

Among the mothers who were pregnant at the time of the nuclear explosions in Japan, and who received sufficiently large doses to show the usual acute radiation symptoms, there was a marked *increase* over the normal number of stillbirths and deaths of infants within a year of birth! The increase in mortality was significant only when the mother had been exposed during the last three months of pregnancy. Among the surviving children there was a slight increase in frequency of mental retardation, and head circumferences were smaller than normal. Radiation had attacked bone and marrow! Peru has found D.M.S.O. is *very* effective in treating their mentally retarded. Irradiated bodies might be neutralized or hopeless.

Maldevelopment of Teeth

This was attributed to injury of the roots, and was also noted in many of the children. (The wolf shall dwell with the lamb, and the leopard shall lie down with the kid; and the calf and the young lion and the fatling together; and a little child shall lead them. The cow and the bear shall feed; their young shall lie down together: and the lion shall eat straw like the ox. The sucking child shall play on the hole of the asp. . . . They shall not hurt nor destroy . . ." [Isaiah 11:6-9]) The fear expressed at one time that there would be a sharp increase in the occurrence of abnormalities had not been substantiated at *that* time. However, *they* know now, and so do the Marshallese. *They* again showed absolute indifference to the welfare of people. The tests must go on! Science must have its guinea pigs, even if they are people!

The Japanese Experience

Early fallout accompanying the nuclear air bursts over Japan were thought to be insignificant and were not monitored for a very good reason: *We were not welcome there at that time!* Consequently, no informaiton was available concerning the potentialities of fission products and other weapon residues as internal sources of radiation. That data have been coming to light over the years since. (Our government has taken the stiff-necked position of flatly refusing to give compensation or aid of any kind to veterans who obviously are suffering from radiation poisoning, notably the clean-up crews assigned to Nagasaki and Hiroshima! *They,* the weapons testers, have applied for and received permission to expose our troops at a closer range of the explosion of atomic weapons than is considered safe by military medical personnel; this was confirmed by the Veterans Administration.)

All pictured exposures on the following pages are of doses of 200 to 1,000 rems; survival is possible for those with doses at the lower end of the scale but unlikely at the upper end. Five thousand rems or more result in prompt changes in the nervous system!

The initial symptoms are similar to those common in radiation sickness, namely, nausea, vomiting, diarrhea, loss of appetite, and a go-to-hell, devil-may-care attitude, which if they know what is happening may be a blessing! The larger the dose, the sooner will these symptoms develop, generally during the initial day of exposure. After the first day or two the symptoms disappear and there may be a latent period of several days to two weeks during which the person or animal feels relatively well, although important changes are occurring in the *blood!* Subsequently, there is a return of all the symptoms and a steplike rise in temperature, which may be due to infection. This may be diagnosed as the flu!

Loss of hair, which is a prominent consequence of radiation exposure, also starts after about two weeks, following the latent period for doses over 300 rems (see Fig. 34).

Internal Hazards

Wherever fallout occurs there is a chance that radioactive material will enter the body through the digestive tract, due to

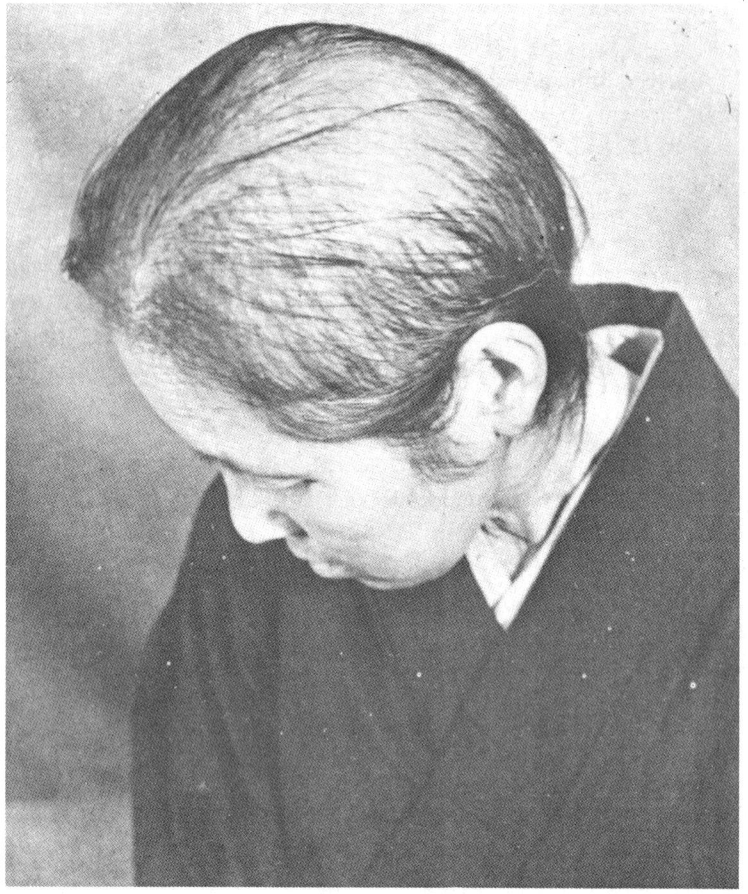

FIGURE 34

An example of loss of hair due to radiation exposure (*From* ENW)

the consumption of food and water contaminated with fission or fusion products (we had better waste no time in roofing our reservoirs) or through the lungs, by breathing air containing fallout particles, or through wounds or abrasions from being tossed around or slammed against objects. Even a very small quantity of radioactive material, if retained in the body, can produce considerable injury. Radiation exposure of various organs and tissues from internal sources is *continuous,* subject only to depletion of the quantity of active material in the body as a result of radioactive decay and biological elimination processes. Furthermore, internal sources of alpha emitters from plutonium or beta particles, or soft, low-energy gamma-ray emitters can deposit a radiation dose of 700 rads/rems over a period of ninety-six hours, which would probably prove fatal in the majority of cases! (See Table 13.)

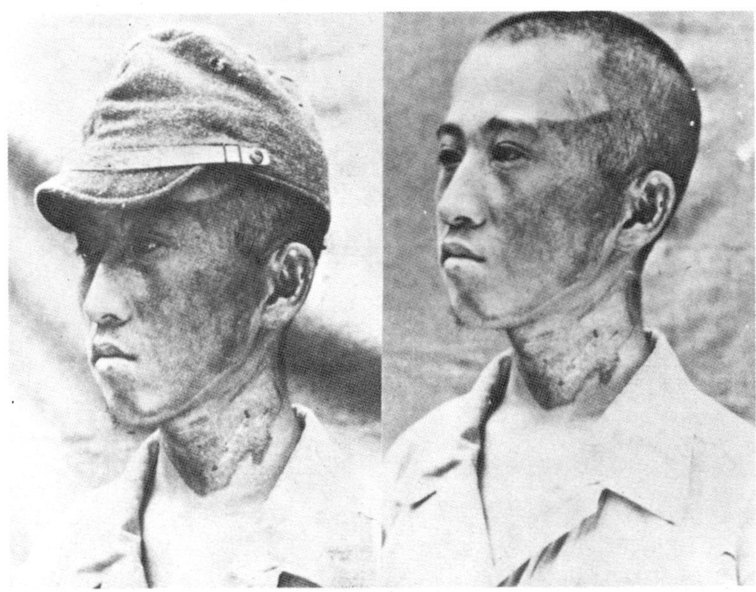

FIGURE 35

Partial protection produced "profile" burns. The cap was sufficient to protect the top of the head against flash burns. (*From* ENW)

External Hazards

Although flash burns were largely confined to exposed parts of the body, there were a few cases where such burns occurred through one, and very occasionally more, layers of clothing. Instances of this kind were observed when the radiant exposure was large enough to overcome the protective effect of the particular fabric. Wear a hat whenever you go out!

There were many instances in which burns occurred through black clothing but not through white material worn by the same individual. This was attributed to the reflection of thermal radiation by white or other light-colored fabrics.

TABLE 13

Summary of Clinical Effects of Acute Ionizing Radiation Doses

Range	0 to 100 rems Subclinical range	100 to 1,000 rems Therapeutic range			Over 1,000 rems Lethal range	
		100 to 200 rems	200 to 600 rems	600 to 1,000 rems	1,000 to 5,000 rems	Over 5,000 rems
		Clinical surveillance	Therapy effective	Therapy promising	Therapy palliative	
Incidence of vomiting	None	100 rems: infrequent 200 rems: common	300 rems: 100%	100%	100%	
Initial Phase Onset Duration	 — —	 3 to 6 hours ≤ 1 day	 ½ to 6 hours 1 to 2 days	 ¼ to ½ hour ≤ 2 days	 5 to 30 minutes ≤ 1 day	Almost immediately**
Latent Phase Onset Duration	 — —	 ≤ 1 day ≤ 2 weeks	 1 to 2 days 1 to 4 weeks	 ≤ 2 days 5 to 10 days	 ≤ 1 day* 0 th 7 days*	Almost immediately**
Final Phase Onset Duration	 — —	 10 to 14 days 4 weeks	 1 to 4 weeks 1 to 8 weeks	 5 to 10 days 1 to 4 weeks	 0 to 10 days 2 to 10 days	Almost immediately**
Leading organ	Hematopoietic tissue				Gastrointestinal tract	Central nervous system
Characteristic signs	None below 50 rems	Moderate leukopenia	Severe leukopenia; purpura; hemorrhage; infection. Epilation above 300 rems.		Diarrhea; fever; disturbance of electrolyte balance.	Convulsions; tremor; ataxia; lethargy.
Critical period post-exposure	—	—	1 to 6 weeks		2 to 14 days	1 to 48 hours

Therapy	Reassurance	Reassurance; hematologic surveillance.	Blood transfusion; antibiotics.	Consider bone marrow transplantation.	Maintenance of electrolyte balance.	Sedatives
Prognosis	Excellent	Excellent	Guarded	Guarded	Hopeless	
Convalescent period Incidence of death	None None	Several weeks None	1 to 12 months 0 to 90%	Long 90 to 100%	— 100%	
Death occurs within	—	—	2 to 12 weeks	1 to 6 weeks	2 to 14 days	< 1 day to 2 days
Cause of death	—	—	Hemorrhage; infection		Circulatory collapse	Respiratory failure; brain edema.

*At the higher doses within this range there may be no latent phase.

**Initial phase merges into final phase, death usually occurring from a few hours to about 2 days; this chronology is possibly interrupted by a very short latent phase.

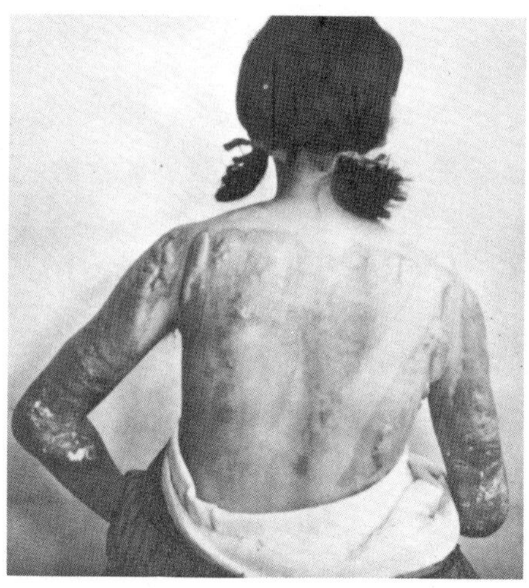

FIGURE 36

Skin under a light layer of clothing was burned. The areas not burned were protected by thicker layers. (*From* ENW)

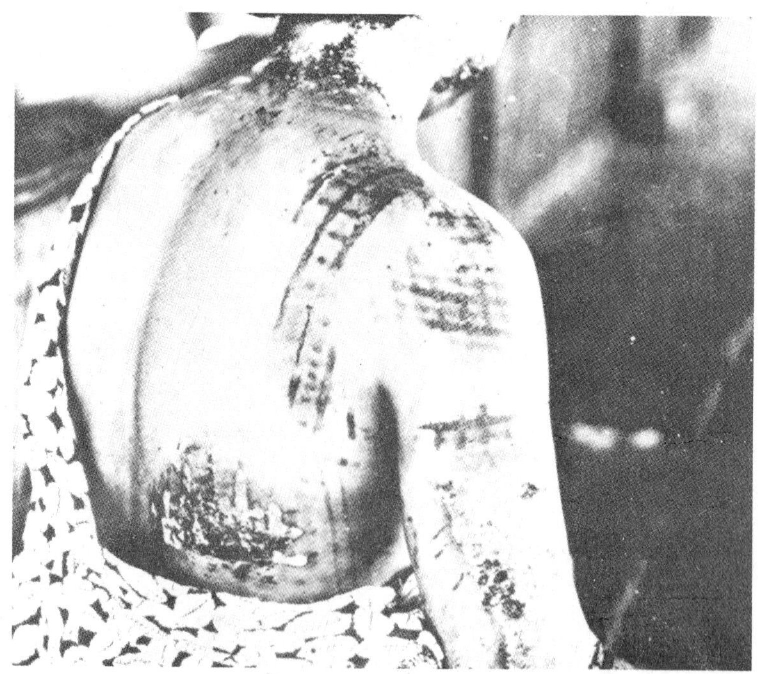

FIGURE 37

Burns correspond to pattern of dark portion of a kimono worn at the time of the explosion. (*From* ENW)

The *ENW* report states:

> The first evidence of skin damage was increased pigmentation, in the form of dark colored patches and raised areas (macules, papules, and raised plaques). These lesions developed on the exposed parts of the body not protected by clothing, and occurred usually in the following order: scalp (with epilation), neck, shoulders, depressions in the forearm, feet, limbs, and trunk. Epilation and lesions of the scalp, neck, and foot were most frequently observed.
>
> In addition, a bluish-brown pigmentation of the fingernails was very common among . . . American Negroes who were in a group of servicemen stationed on Rongerik Atoll. The phenomenon appears to be a radiation response peculiar to the dark-skinned races, since it was not apparent in any of the white Americans who were exposed at the same time. The nail pigmentation occurred in a number of individuals who did not have skin lesions. It is probable that this was caused by gamma rays, rather than by beta particles, as the same effect has been observed in dark-skinned patients undergoing X-ray treatment in clinical practice. . . .
>
> Individuals who had been more highly contaminated developed deeper lesions, usually on the feet or neck, accompanied by mild burning, itching, and pain. These lesions were wet, weeping, and ulcerated, becoming covered by a hard, dry scab; however, the majority healed readily with the regular treatment generally employed for other skin lesions not connected with radiation. Abnormal pigmentation effects persisted for some time, and in several cases about a year elapsed before the normal (darkish) skin coloration was restored.
>
> Regrowth of hair, of the usual color (in contrast to the skin pigmentation) and texture, began about nine weeks after contamination by fallout and was complete in six months. By the same time, nail discoloration had grown out in all but a few individuals.

Pain associated with skin burns occurs when the temperature of certain nerve cells near the surface is raised to 43°C. (109°F.) or more. That is what you can expect if you are subjected to early fallout, in addition to inhalation problems! There is really nothing to fear for humanity in any part of the world but a slow, agonizing death for *all,* in case of a nuclear war!

Whether you are involved or not, all of humanity will suffer and pray for a quick death. The fire from the skies has already enclosed the earth in a deadly shell. What a gas chamber that will be! Of course, it could be God's way of isolating the fire from the sky, for a nice even burn all over the earth.

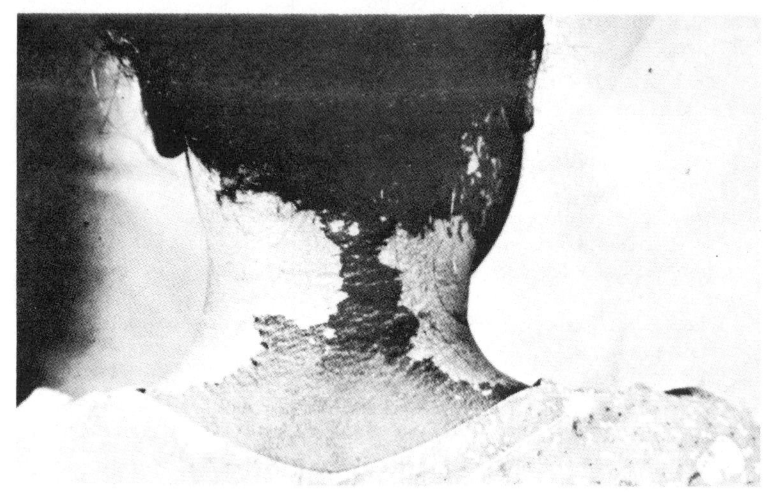

FIGURE 38

Beta burn on neck one month after exposure. (*From* ENW)

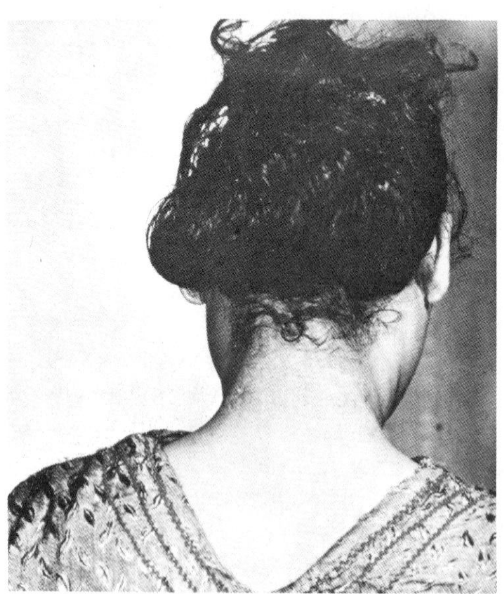

FIGURE 39
Beta burn on neck one year after exposure. (*From* ENW)

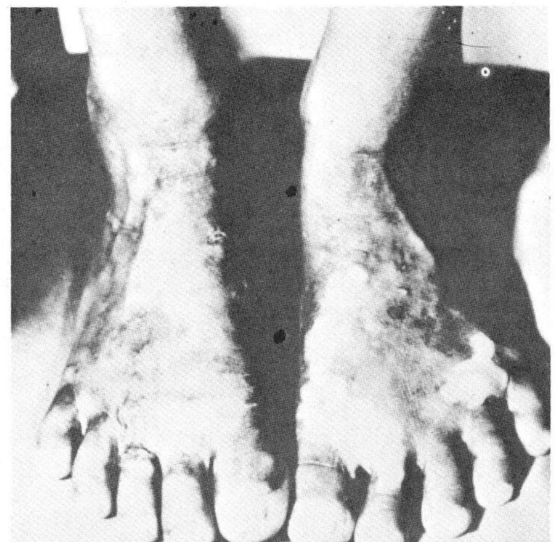

FIGURE 40

Beta burn on feet one month after exposure. (*From* ENW)

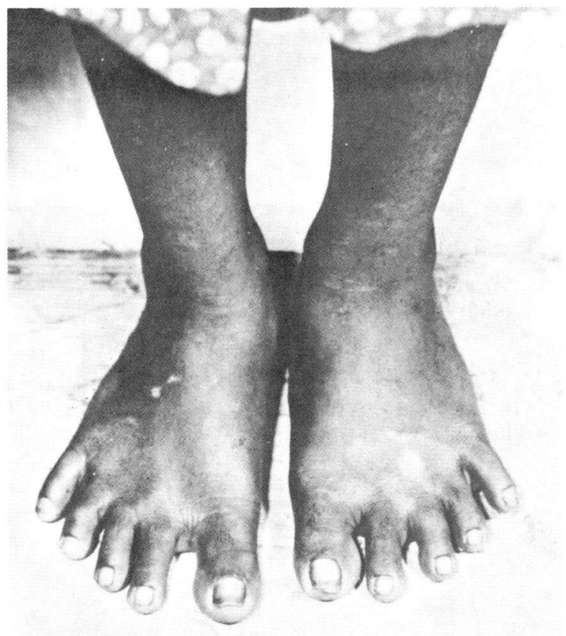

FIGURE 41

Beta burn on feet six months after exposure. (*From* ENW)

LATE EFFECTS OF RADIATION

A number of consequences of nuclear radiation may not appear for some years after exposure, regardless of the short biological half life of certain radioactive elements or rapid passage through and elimination by the body! Among these consequences, apart from genetic effects, are the formation of cataracts, nonspecified causes of life-shortening, leukemia, other forms of malignant disease, and retarded development of children *in utero* at the time of exposure. Information concerning these late effects has been obtained from continuing studies, including those in Japan, made chiefly under the direction of the Atomic Bomb Casualty Commission.

The problems that arise later in life, like the acute reactions observed shortly—a few weeks or months—after irradiation come from changes in cells and tissues at the time of exposure. If a person survives the acute reaction of irradiation, damaged cells may be replaced but not be completely normal; however, the causes for effects that occur later are largely unknown, although there are many theories.

Cataracts

The term "cataract" is commonly used to describe any detectable change in the normal transparency of the lens of the eye. Sufficient exposure of the eye lens to ionizing radiations can cause cataracts in man and in other animal species. There is a latent period between exposure of the eye and development of an opacity ranging from a few months to several years. The duration of the latent period is shorter the greater the radiation exposure. Fast neutrons, such as those from a nuclear explosion, have been found to be more effective than gamma rays in producing cataracts!

Leukemia

Mortality rates show that prior to about 1950 leukemia, a form of cancer in which there is a proliferation of white cells in the blood, was a much more common cause of death of radiologists than of other physicians. This has been attributed to chronic exposure to moderate amounts of radiation over many years.

After protective measures were improved, the incidence of leukemia among radiologists declined markedly. A larger than expected frequency of leukemia has also been observed among a group of people in the United Kingdom who had received large doses of X rays to the spine for the treatment of a form of arthritis known as ankylosing spondylitis. Three main types of leukemia are induced by radiation, namely, acute and chronic granulocytic and acute lymphocytic forms; the occurrence of chronic lymphocytic leukemia is not significantly increased by radiation. The development of leukemia as a result of an overexposure to radiation is associated with a latent period varying from one to twenty years or more. The disease is generally fatal, no matter what its cause!

Other Types of Cancer

Mortality statistics show that exposures to radiation for various medical purposes can increase the frequency of various types of cancer, in addition to leukemia. The same effect has been observed among the survivors of the nuclear attacks on Japan. For example, after a latent period of about ten years, a significant increase was observed in the incidence of thyroid cancer among individuals who were within about half a mile from ground zero and consequently received large doses of ionizing radiations.

PEOPLE-EATER ELEMENTS

Alpha and beta particles are practically harmless, *unless* in contact with the body or ingested. Iodine-131 has a half life of 8 days. Iodine-132, 133, and 134 have half lives of less than a day. The biological half life in the thyroid is 80 days for most, years for others!

The biological half lives of strontium-89 and 90 are 27.7 years. It is *very* dangerous!

The biological half life of cesium-137 is 30 years! It is also very dangerous.

The biological half life of barium-140 is 12.8 days; also *very* dangerous.

The rare-earth elements, ceruim-144 and yttrium-91, also must be considered dangerous.

Plutonium-239, is an alpha-particle emitting isotope and has a long radioactive half life of 24,000 years! It also has a long biological half life of about 100 years in the *skeleton,* and 40 years in the *liver!* It may be present in the early fallout from the bombs. As with any inhaled fallout, particles contaminated with plutonium will be deposited in the lungs.

Recent research offers hope that the disastrous effects of plutonium may be eliminated, according to the article reprinted below:

'CURE' FOUND FOR PLUTONIUM

By Paul Raeburn

BERKELEY, Calif. (AP) — A substance that removes deadly plutonium from living tissue has been developed by University of California scientists, who call it the first advance in treating radiation contamination in more than 30 years.

It also could prove valuable in removing radioactive elements from nuclear waste, thus defusing the highly controversial problems of storing spent plutonium, as well as low-level wastes from nuclear medicine and research, the scientists say.

"All of the chemistry for reactor-waste storage was developed shortly after the Manhattan Project," the intensive World War II effort to develop the atomic bomb, said Kenneth Raymond, one of the substance's inventors.

"I think this new chemical, or something very close to it, will prove to be a significant part of the answer" for disposal of wastes from nuclear power plants and weapons, he said.

The substance, called LICAM-C, removed 70 percent of the plutonium injected into laboratory mice. And repeated doses could probably remove more with little or no toxic side effect, researchers said.

Plutonium, one of the deadliest substances known, is both a byproduct of and a fuel for nuclear power plants and weapons manufacture. Chemically similar to iron, it is easily absorbed by the body and collects in the lungs, spleen, liver and bone barrow. The radiation it emits can transform normal cells into cancerous ones.

Plutonium-239, a nuclear fission fuel, has a half-life of 25,000 years, which means half the plutonium will decay in that time. That enormous half-life and plutonium's deadly radiation are at the crux of the storage problem.

Developed by Raymond and Fred Weitl at the University of California's Lawrence Berkeley Laboratory, LICAM-C grabs individual plutonium ions—electrically charged atoms—and engulfs them in chemical "pincers," the researchers said.

"Finally we came up with a chemical that binds tightly

> with plutonium, is non-toxic, and of low enough molecular weight to pass through the kidneys so it can be excreted," Raymond said.
>
> Some earlier experimental substances removed plutonium, but also removed essential minerals such as iron, calcium and zinc, he said. LICAM-C is selective for plutonium.

Now that's good news. Despite the large amounts of radioactive material that may pass through the kidneys in the process of elimination, these organs seem not to be greatly affected by radiation! However, uranium can act as a chemical poison as well, and the quantity of uranium found in fallout is very small. (What the hell, if one don't getcha, the other will.)

Now I have no vendetta with politics or people, but some questions have come to mind regarding statements in the article above. Thank God for the University of California's Lawrence Berkley Laboratory and Raymond and Fred Weitl. I see according to them that plutonium-239 has a half life of 25,000 years instead of the 24,000 most of us have heard about. Oh, well, what's 1,000 years between friends?

Paragraphs two and three of the article imply that nuclear waste can be neutralized; that through nuclear research ways have been found to clean out the radiation used in thyroid scans, instead of leaving it in the thyroid to cause trouble. Thus, radioactive iodine scans whose objective is to destroy part of the gland when it is overactive but carries the threat of radiation can now be a sensible therapeutic measure instead of questionable prophylaxis.

Thank God for Kenneth Raymond, one of the researchers. Why was the ability to do this concealed from the public so long? Why the flap over the problem of storing radioactive waste? Was that why all the labs involved in nuclear research were by administrative order closed down? Thank God it didn't stick. Of course, to cover up, all hydroelectric projects were included, even some that were almost completed. The oil companies were as terrified of new forms of energy as the tobacco people were when the surgeon general forced them to print a warning on every package of cigarettes. Now the tobacco companies own the candy companies. The oil companies now own most of the coal mines and probably some of the electric companies. The only reason we don't have solar power now is that no one can find a place to hang the meter!

Elements other than iodine, strontium, barium, and the rare-earth group were not found to be retained in appreciable amounts in the body—they are merely sojourners passing through. Strontium and barium, which are chemically similar to calcium, are deposited in bone! The radioisotopes of earth elements—cerium metal—which constitutes a considerable portion of fission products, as well as plutonium, are also bone seekers; they can damage bone marrow and cause cancer. They are there to stay!

The amount of radioactive material absorbed from early fallout by inhalation appears to be relatively small, because the nose can filter out almost all particles over 10 micrometers in diameter and about 95 percent of those exceeding 5 micrometers (a micro-inch = 1 millionth of an inch), and that's pretty small! Only a small part of early fallout particles will succeed in reaching the lungs. A wet cloth over your nose might help a lot. Whatever you do don't breathe through your mouth!

Strontium-90

Because of its relatively long radioactive half life of 27.7 years and its appreciable yield in the fission process, strontium accounts for a considerable fraction of the total activity of the fission products that are several years old. Strontium is chemically similar to calcium, an element essential to both plant and animal life; an adult human being, for example, contains over two pounds of calcium, mainly in bone. However, the relationship between strontium and calcium is not a simple one; because of its complex metabolism in the body, the behavior of strontium-90 cannot be stated in terms of a single effective half life.

The probability of serious pathological change in the body of a particular individual, due to the effects of radioisotopes deposited internally, depends upon the amount deposited, the energy of the radiations emitted, and the length of time the source remains in the body. Strontium-90 and its daughter, yttrium-90, emit beta particles which can cause serious localized damage, following their deposition and long-term retention in the skeleton. Tests with animals indicate that the pathological effects resulting from sufficient quanties of inhaled, ingested, or injected strontium-90 include bone necrosis, bone tumors, leukemia, and other hematologic dyscrasias (abnormalities).

As there has been no experience with appreciable quantities

PLATE 10
Are there giants alive today? *(Photo by Ethel Howard)*

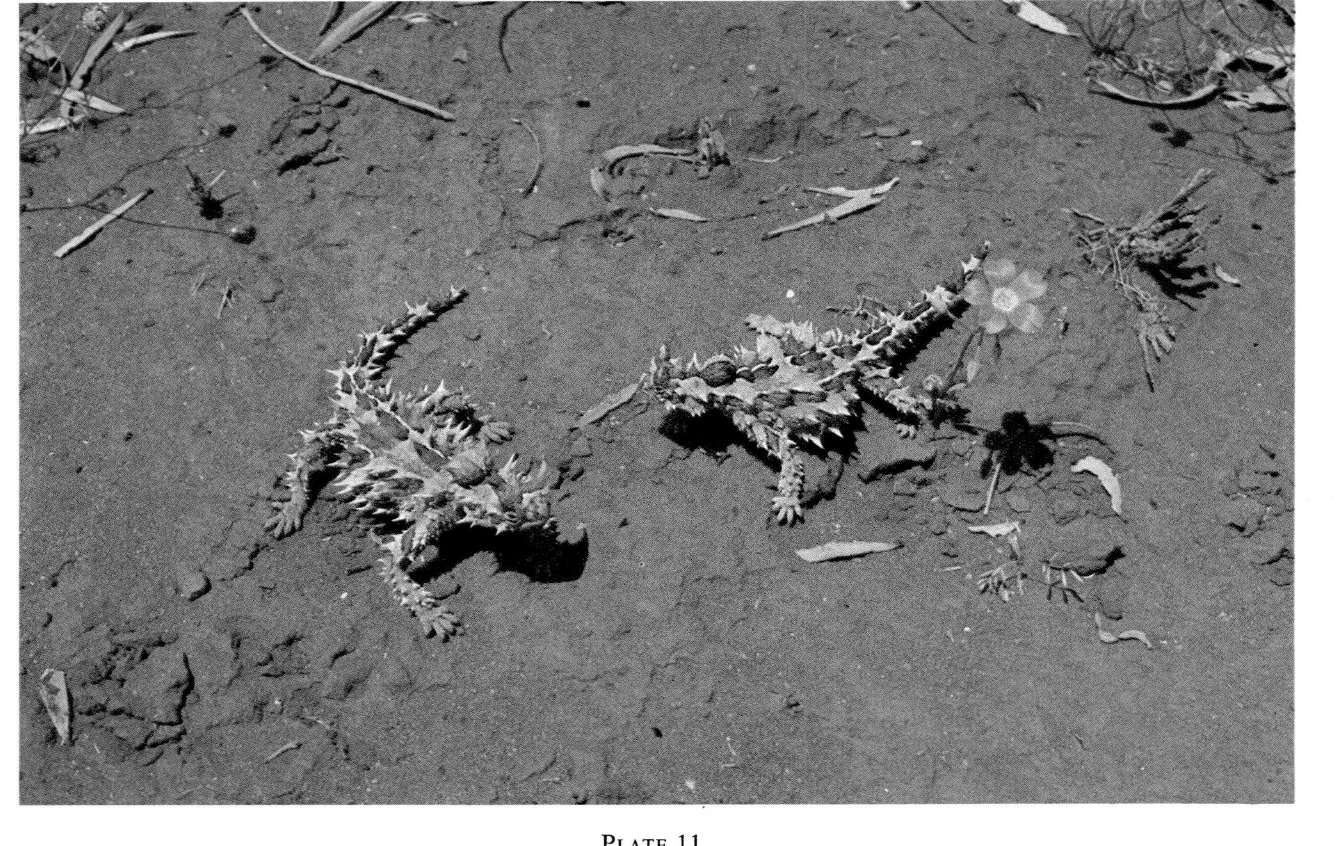

PLATE 11

Might thorny devils—lizards from Australia—be imported to counter the threat from Formosan termites? *(Photo by Hans and Judy Beste, Tom Stack & Associates, Colorado Springs, Colorado)*

PLATE 12

Flash point! Thousands of atomic bombs detonated at once might be enough to boost the earth out of orbit and send it whirling through space a flaming mass! *(Photo makeup by Wy-East Color Laboratory, Portland, Oregon; photo of the earth courtesy NASA)*

Armageddon? *(Lorenzo Ghiglieri artist; mural © 1974*

the Universe, Inc., Dempsey Center, Portland, Oregon)

PLATE 14

Dust storm approaching the Syrian ruins of Dura Europus, built *ca.* 300 B.C.; the fort was abandoned *ca.* A.D. 256. *(Photo courtesy Jean Dufour, Avignon, France).*

We must make up our minds: Do we want a worldwide dust bowl . . .

PLATE 15

. . . or shall we make sure that our earth remains green and beautiful—the way nature intended? *(Photo by permission of Woodfin Camp, Inc.)*

In the meantime, when there were gathered together an innumerable multitude of people, insomuch that they trod one upon another, he began to say unto his disciples first, "Beware ye of the leaven of the Pharisees, which is hypocrisy.

"For there is nothing covered that shall not be revealed; neither hid that shall not be known.

"Therefore whatsoever ye have spoken in darkness shall be heard in the light, and that which ye have spoken in the ear in closets shall be proclaimed upon the housetops.

"And I say unto you my friends, Be not afraid of them that kill the body, and after that have no more that they can do.

"But I will forewarn you whom ye shall fear: Fear him which, after he hath killed, hath power to cast into hell; yea, I say unto you, fear him!"

And he said also to the people, "When ye see a cloud rise out of the west, straightway ye say, 'There cometh a shower'; and so it is.

"And when ye see the south wind blow, ye say, 'There will be heat'; and it cometh to pass.

"Ye hypocrites, ye can discern the face of the sky and of the earth, but how is it that ye do not discern this time?"

Luke 12:1-5, 54-55.

of strontium-90 in the human body, the relationship between the probability of serious biological effect and the body burden of this isotope is not known with certainty. Tentative conclusions have been based on a comparison of the effects of strontium-90, with radium on test animals, and on the known effects of radium on human beings. From these comparisons it has been estimated that a body content of 10 microcuries (one microcurie is one-millionth part of a curie) of strontium-90 in a large proportion of the population would produce a noticeable increase in the occurrence of bone cancer. On this basis, it has been recommended that the maximum activity of strontium-90 in the body of any individual who is exposed in the course of his occupation be taken as 2 microcuries. Since the average amount of calcium in the skeleton of an adult human is about 1 kilogram (or a little over 2 pounds), this corresponds to a concentration in the skeleton of 2 microcuries of strontium-90 per kilogram of calcium. Moreover, the limit generally considered to be acceptable for any individual member of the general population is 0.067 microcurie per kilogram of calcium!

As a result of nuclear test explosions in the atmosphere by various countries, there has been an increase in the strontium-90 content of the soil, plants, and the bones of animals and man. *It may be already later than we think. It may be too late!* This increase is *worldwide* and is not restricted to areas in the vicinity of the test sites, although it is naturally somewhat higher in these regions because of the proximity to the detonation (fast fallout). The fine particles of the delayed fallout descend from the stratosphere into the troposphere over a period of *years,* and are then brought down by rain and snow. Consequently, the amount of strontium-90 in the stratosphere available to fall on earth is determined by the difference between the quantity and radioactive decay. This net amount reached a maximum at the end of 1962, after the cessation of nuclear weapons testing in the atmosphere by the United States and the USSR (see Fig. 42). Is it a question of old too soon and smart too late?

Carbon-14

Carbon-14 does not tend to concentrate in any particular part of the body and is distributed almost uniformly throughout soft tissue; hence, the whole body is exposed to the low-energy

Beta particles. The whole-body dose from carbon-14, in nature before 1952, was somewhat less than 1 millirem per annum. By 1964, this dose had been roughly doubled by the additional carbon-14 arising from nuclear tests in the atmosphere. If there are no further substantial additions, the dose will decrease gradually and approach normal in another 100 years or so. Compared with the annual radiation dose from strontium-90, mainly to the skeleton, the contribution from carbon-14 produced by thermonuclear weapons is small.

GENETIC EFFECTS OF NUCLEAR RADIATION

The harmful effects of a deleterious mutation may be moderate, such as increased susceptibility to disease or a decrease in life expectancy by a few months, or they may be more serious, such as death in the embryonic state. Thus, individuals bearing harmful genes are handicapped relative to the rest of the population, particularly in the respects that they tend to have fewer children or die earlier. It is apparent, therefore, that a person with such genes will be eliminated from the population. A gene that does great harm will be eliminated rapidly, since few (if any) individuals carrying such genes will survive to the age of reproduction! On the other hand, a slightly deleterious mutant gene may *persist* much longer and, thereby, do harm, although of a less severe character, to a larger number of individuals.

Another matter of interest in connection with the effects of nuclear explosions on plants and farm animals is the possibility of *serious ecological disturbances*. These might be caused by large-scale fires, denuding of forests by fallout, destructive plagues of insects that are known to be relatively insensitive to radiation, and so on, until the seven plagues have occurred!

The graphs shown in Figs. 42 and 43 show the variations over a period of years of the megacuries of strontium-90 present in the total stratospheric area, i.e., the activity still remaining in the stratosphere at various times, and the ground inventory i.e., the activity deposited on the ground. The extensive atmospheric nuclear test programs conducted by the United States and the USSR during 1961 and 1962 are reflected by the large amount in the stratospheric area, which reached a maximum toward the end of 1962. The sharp increase in the ground inventory, which began in 1962 and continued through 1965, reflects the deposition of strontium-90 during those years.

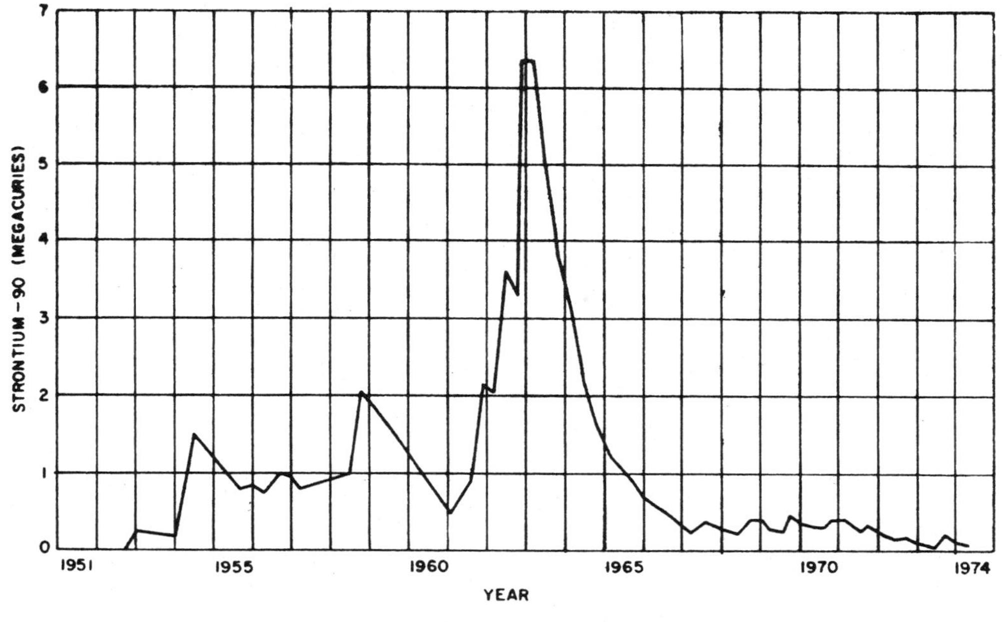

FIGURE 42

Stratospheric inventory of strontium-90. (*From* ENW)

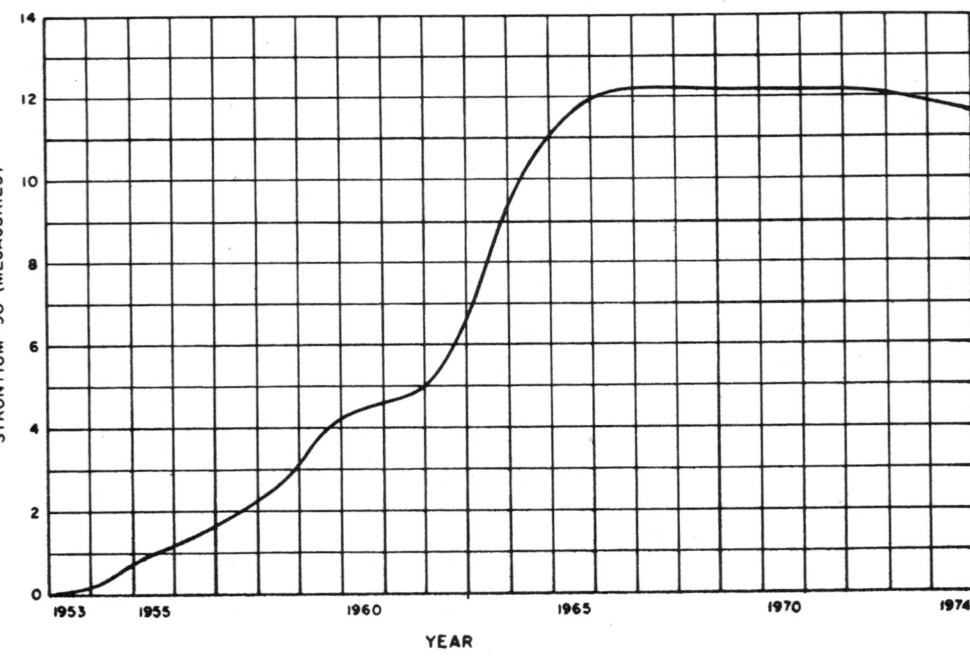

FIGURE 43

Surface inventory of strontium-90. (*From* ENW)

The maximum amounts of strontium-90 on the earth's surface will be attained when the rate of natural radioactive decay just begins to exceed the rate at which the isotope reaches the ground in delayed fallout. The atmospheric tests conducted by France, India, and China, in the late 1960s and early 1970s, did not do us any good. Pressure must be brought on them to cease all of the atmospheric testing! If it were discontinued at once, the surface loading would begin to decrease at once—it's *URGENT!*

6

The Structure of the Atmosphere

One of the most significant discoveries about the atmosphere is the variation in temperature at different altitudes and its dependence on latitude and time. Again it looks like *they* blew it! No mention as to how this information was discovered or when seems to be available. It was long assumed that as we got further away from earth, the atmosphere became a vacuum with absolute zero. Not so, my friends, it's a fiery furnace out there! In *ascending* into the atmosphere from the surface of the earth, the temperature of the air falls steadily, in general, toward a minimum value. This region of falling temperature is called the "troposphere" and its top, where the temperature ceases to decrease, is known as the "tropopause." Above the troposphere is the stratosphere, where the temperature remains more or less constant with increasing altitude in the temperate and polar zones. Although all the atmosphere immediately over the tropopause is commonly referred to as the stratosphere, there are areas in which the structure varies (see Fig. 44). In the equatorial regions, the temperature in the stratosphere increases with height! This inversion also occurs at the higher altitudes in the temperate and polar regions! In the "mesosphere" the temperature falls off again, with increasing height. At still higher altitudes is the "thermosphere" where the temperature rises rapidly with height!

Most of the visible phenomena associated with weather occur in the troposphere (see chapter 1). The high moisture content, the relatively high temperature at the earth's surface, and the convective movement (or instability) of the air arising from temperature differences promote the formation of clouds and rainfall. In the temperate latitudes, at about 45 degrees in the summer and 30 degrees in the winter, where the cold polar air meets the

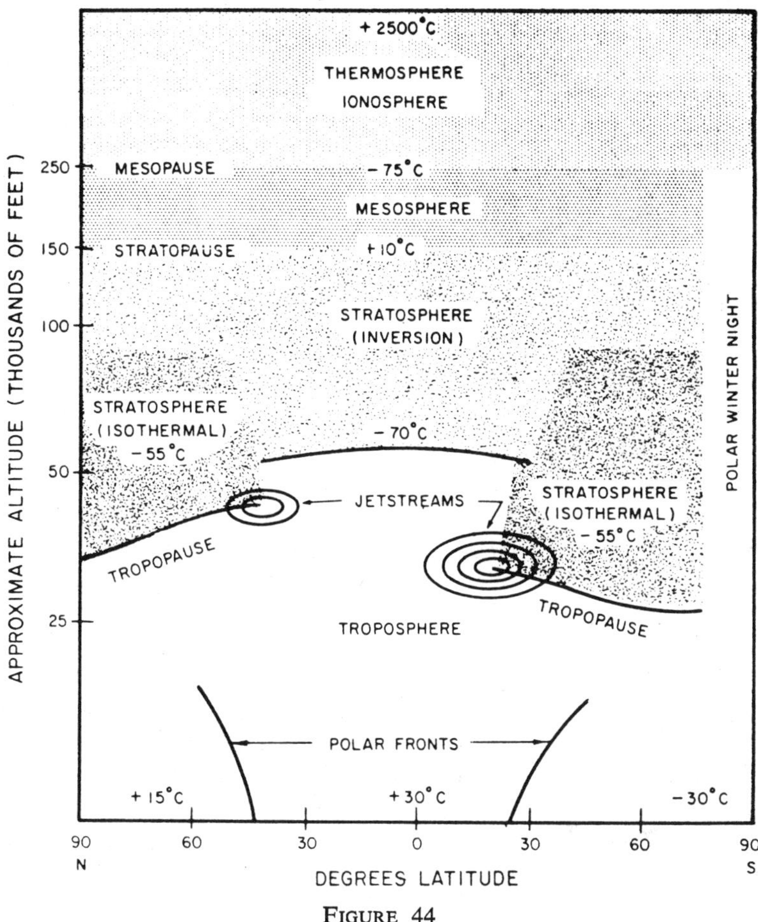

FIGURE 44

Structure of the atmosphere during July and August. (*From* ENW)

warm air of the tropics, there are formed meandering, wavelike bands of storm fronts called "polar fronts" (see Fig. 44). In these regions, the average rainfall is high.

The tropopause (top of the troposphere) is lower in the polar and temperate zones than in the tropics; its height in the former regions varies from 25,000 to 45,000 feet, depending on latitude, time of year, and particular conditions of the day. In general, the altitude is lowest in the polar regions. The tropopause may disappear entirely at times in the polar winter night. In the tropics, the tropopause usually occurs near 55,000 feet at all seasons! It is more sharply defined than in the temperate and polar regions because in the tropics the temperature increases with height above the tropopause instead of remaining constant. There is a marked gap or discontinuity in the tropopause in each temperate zone, as may be seen in Fig. 44, that constitutes a region of unusual turbulence. Each gap moves north and south seasonally, following the sun, and is usually located near a polar front. It is believed that considerable interchange of air between the stratosphere and troposphere takes place at the gaps. A jet stream, forming a river of air moving with high speed and circulating about the earth, is located at the tropical edge of the polar tropopause in each hemisphere.

Because of its temperature structure, there is very little convective motion in the stratosphere, and the air is exceptionally stable. This is especially noticeable in the tropics where the vertical movement of the radioactive cloud from a nuclear explosion has sometimes been less than two miles in three trips around the globe (about 70,000 miles). This stability continues up to the mesosphere where marked turbulence is again noted. The polar stratosphere is less stable than that in the tropics, particularly during the polar winter night when the stratospheric temperature structure changes to such an extent that the inversion may disappear. When this occurs there may be considerable convective mixing of the air to great heights!

Many readers may recall an airline steward announcing the fact that the aircraft is flying at 35,000 feet and the temperature outside is −55 degrees. (Who cared whether it was Centigrade or Fahrenheit; no one intended to step outside and see.) However, now you know just about where you were in the atmosphere, and you could expect the temperature to continue to drop to −70°C. If you are in a Concorde on a trip to London, you can expect things to warm up to +5°C. or 10°C., though the plane rarely flies over 70,000 feet. At +10°C. you would find yourself almost out of the stratopause and entering the mesosphere, where the old thermometer starts a nose dive and at 250,000 feet hits a low of −75° C. Then you would be in moon rocket territory, entering the ionosphere and thermosphere, where it would blow out the thermometer at +2,500° C.

This structured atmosphere was discovered by moon rocket instrumentation, which took place after the advent of the atmospheric bomb testing. In any event, a hole was blown in the tropopause and apparently the debris was neatly folded back over the surface of the troposphere. The jet streams were instantly

Rough weather racks world

By United Press International

China and other parts of Asia are suffering a searing heat wave similar to the one that gripped much of the United States earlier this month. But in much of Europe, summer arrived late.

Weather officials said Peking is suffering its driest period in 100 years and already the city's 84 reservoirs are virtually dry. City and commune authorities have been holding emergency meetings to try to save crops.

But in a summer of unpredictability, Typhoon Joe last week slammed into southern China with 100 mph winds, killing 140 people, causing extensive flooding and cutting communications.

Taiwan has been reeling under average daily temperatures of 93 degrees the past three months.

In India, monsoon rains may be too late to save this year's crops. Despite more than 75 million acres of well-irrigated farmland, most of India remains at the mercy of the weather. "It is a case of receiving too much or too little rain," said J.C. Pant, a New Delhi-based

homes and left nearly half a million homeless.

In Europe, unseasonable cold, rain and storms have left a trail of ruined crops and vacations across the continent during the past two months.

But warm sunshine and gentle breezes are at last tempting people out of their raincoats and into their swimsuits but pockets of win-

Melanoma cases puzzle Livermore Lab officials

LIVERMORE, Calif. (AP) — A total of 27 employees at the Lawrence **Livermore Laboratory have contracted the rare skin cancer melanoma** since the nuclear weapons research facility opened in 1952, the lab's medical director reported Friday.

Dr. Max Biggs said 18 **cases of malignant melanoma were found between 1972 and 1977, compared with just nine cases in the previous 20 years, and officials were puzzled by the sudden dramatic increase in the last decade.**

According to Dr. Donald Austin of the state Department of Health Services, **lab employees apparently were exposed to some as yet unfound cancer-causing agent in the 1970s.** Austin said there was no doubt it **was an environmental carcinogen.**

"We know from past experience that some cancer rates have changed dramatically from exposure to such a carcinogen," he said.

Briggs said that throughout the United States in the past five or six years "the incidence rate has doubled, so there is a nationwide increase in this tumor for some unexplained reason."

In a report made public last month, Austin said the rates at the lab were three to five times higher than in the local population outside the lab.

Briggs said no common denominator was evident in any of the 27 cases of melanoma found among employees of the lab.

Four sufferers had less than four years of service at the lab when their cancers were diagnosed; five had 20 years. **The work of the stricken employees ranged from gardening to physics, Briggs said.**

Melanoma is believed to be caused in many cases by overexposure to ultraviolet radiation, such as sunlight.

Severe lightning storms lash broiling Southwest

By United Press International

The hot, dry weather that has broiled been reported from the Mexican border to Palm Springs," he said. "We are seeing a

Tornadoes, hail batter Southwest

By United Press International

Storms spiked with tornadoes, flooding and grapefruit-sized hail battered parts of Oklahoma and Texas Thursday, causing at least $2 million in property damage.

The storms pushed eastward into the Midwest and Southeast, knocking out power and telephone service in Chicago suburbs and flooding streets in Alabama.

A gas well hit by a tornado in Reydon, Okla., spewed natural gas for several hours late Wednesday until workers could close it off. At least 15 homes and 10 barns were damaged severely by the same tornado.

Officials estimated damage in the Reydon area at $2 million. No serious injuries were reported.

Power and telephone lines near Cheyenne, Okla., were knocked out by another tornado late Wednesday that was accompanied by grapefruit-sized hail and about 2 inches of rain.

watches through the night and morning.

Tornadoes swept the Texas towns of Tulia, Nazareth and Allison Wednesday night, damaging barns, garages and trailers and injuring three oil derrick workers.

The National Weather Service reported that six twisters touched down in eastern North Dakota Wednesday.

Wind gusts up to 85 mph raked Langdon, N.D., and large hail pounded other sections of the state.

The tornado at Allison, in the northeast Panhandle, destroyed at least four mobile homes and knocked three workers from a nearby drilling rig. The men suffered back injuries but were reported in good condition.

Hail the size of softballs bombarded southern Roberts County in Texas. Pampa, Texas, reported hail the size of tennis balls and small hail accumulated 6 inches deep near Levelland, Texas.

Thunderstorms Tuesday and Wednesday brought severe flooding to parts of

Clouds black out London

A storm barreled into Lagoda Monday killing Twailia G. Emery, demolishing her mobile home. H. Clint Harrison, was injured.

"I've never seen anything like t fore," said Sue Dickerson, town treasurer at Ladoga. "The wind w rific. There was no warning. It w here." At least 18 persons were tre local hospitals for tornado-related in

Six persons were injured, sco windows blown out and roofs ripped off

LONDON (UPI) — Freak clouds 6 miles thick blacked out London during the weekend and brought the worst July thunderstorms in a decade. "I've never known it so black in day time," said a London Weather Center official of Saturday when the center's sunlight meter dropped to a reading of zero. During the heavy rain, power supplies to some villages were cut off, roads were flooded and two women golfers were hospitalized with injuries from a lightning bolt.

Connecticut reeling from deadly tornado

WINDSOR LOCKS, Conn. (UPI) — A shaken Gov. Ella Grasso took a dawn flight Thursday over a tornado-ravaged section of her state only a mile from her home, then said she will ask President Carter to declare it a major disaster area.

"It's every bit as bad as we thought," she said after her tour in one of the few Air National Guard helicopters operable after the sudden Wednesday storm that ransacked Bradley International Airport and a nearby business and residential area, killing at least one person.

"I've never seen anything this bad," she said.

"The storm struck without notice. When you see the devastation, you marvel that it was contained even to the extent that it was. This certainly is a tragedy of considerable proportions."

The tornado struck so close to the National Weather Service station at the airport that there was no warning other than darkening skies Wednesday afternoon.

Within moments, one person was dead and 418 others were injured, said State Health Commissioner Douglas Lloyd.

He said 300 persons released at emergency area hospitals. Another hospitalized, including tion.

A search was under

believed to have been in her home when the storm struck.

The governor estimated the tornado caused more than $214 million damage to the business-residential area along the east side of the airport, which sustained about $100 million in damage itself.

The Connecticut Air National Guard reported about $50 million damage to its aircraft.

Winds measuring as high as 86 mph flattened buildings and tossed heavy airplanes and trucks around like paper throughout a 4-square-mile area less than a mile from the governor's home. She said more than 125 buildings were damaged.

The governor said Windsor Locks sustained $100 million damage and that 1,800 persons will be out of work "for a protracted period of time."

Northeast Utilities said power was still out to 2,100 persons and could remain off as long as two days. The state's tobacco crop, stored in barns in the Suffield area, was "extensively damaged."

There was no estimate of the damage done to the Bradley Air Museum, where one of the world's largest aircraft

Rain

FIGURE 45

Is there any doubt that "they blew our weather"? This montage of news stories should help prove it.

lost to all commercial aircraft! Now the U.S. Weather Service looks for them and notifies them of their whereabouts; it is possible to cut the fuel burn about in half on overseas flights if the jet streams can be located. There seems to be three main jet streams (the east to west equatorial does not circle the earth): the subtropic west to east, the polar west to east, and the polar night jet in the stratosphere in winter only. These travel 200 mph plus. No wonder Pan American and other commercial lines like to fly the polar routes!

It is unnatural for rings of particles whether captured from the solar winds as the Van Allen radiation belts (1 and 2) or blown out from the earth or from the earth's atmosphere to form an incomplete ring or shell around this old globe. Nothing else does! If in some manner the debris from the troposphere, stratosphere, mesosphere, and thermosphere, after having been blown clear out of the known structure of our atmosphere, could settle, all of it, back into the crater created, things still wouldn't be the same. Remember the fountain effect! Being lighter by far than the soil we tread on daily, what settled into the sky crater would make a mighty poor shield against the ultraviolet rays continuously emitted by the sun. What about the extreme heat and the hundreds who might die, as the gap, or big window moving north and south following the sun seasonally, lets through the many deadly rays?

The big window—or ditch or gap—might be likened to a drain in a floor or sink, sucking down highly radioactive particulates that would have stayed up in the stratosphere until gravity pulled them down; by that time, their half life would have been spent and they would be more or less harmless to life on earth.

Other planets have particulate rings around them, and they are in perfect circles. Saturn's rings are 41,500 miles wide; Jupiter and Uranus also have perfect rings. Of course the planetary rings are flat and not part of the atmospheric makeup. However, the universal laws of orbiting particles and atmospheres are no doubt the same everywhere. Have we notched the handle of the ray gun aimed at our earth? Have we blown away our bullet-proof vest, the Van Allen radiation belts? Has anyone checked the fields supposedly cooked by the sun with a Geiger counter? A gap, 60° wide, roughly 1,000 miles plus, even if it only cuts *up* into the stratosphere a few miles, worries me!

7

Giantism

ARE THERE GIANTS ALIVE TODAY?

The spelunker standing on the giant's hand swears there are! His wife, hunting with a camera around the Oregon caves, after climbing a steep hill, sat down at the foot of a giant fir tree to catch her breath. The day, with the temperature in the upper 80s, aided by the warm sun filtering through the branches, was perfect for a nap. Sitting there perfectly still with her camera at the ready for whatever wildlife might come by, she fell sound asleep. Awakened by a sharp earthquake and a crashing of brush, she sat ready for a picture of whatever was coming. Most animals have seen people and fear them, but coming upon a person squatting or sitting down, they pause for a second look—and a very good picture! I know—when a wolf that had been sleeping behind a log awakened and saw me squatting just off the trail, he turned his head from one side to the other, then jumped down the trail and loped out of sight. My rifle was leaning against the log under the wolf's nose less than five feet away!

The spelunker, together with a friend, had left his wife early that morning to explore the caves deeper than anyone had ever gone before. Then the earthquake hit, opening a huge crevasse, which the spelunker's buddy dropped into. The crevasse closed instantly, sealing him from the world forever! When the dust settled, the spelunker discovered the cave ceiling had opened high above, flooding the area with light. Standing in front of him was a giant, brushing debris out of his hair and off of his body. He spotted the spelunker at once, reached down and picked him up. The spelunker was already scared out of his wits, thrown to his knees by the quake at the edge of a crevasse, and now this giant!

The giant, all of sixty feet tall, climbed out through the opening and walked into the forest, directly in front of the woman's camera (see Plate 10). He opened his hand allowing the spelunker to stand up. They stared at each other a second or two, then the spelunker looked around and discovered he was sixty feet or more from the ground. The giant stooped down and placed his hand on the ground. Spelunker, 190 pounds and a bit over six feet tall, lit, running! Not a word was spoken. The spelunker, now behind the nearest tree, saw the giant turn and pick up a huge boulder. With a few strides he reached the hole in the ground, climbed down into the cave, plugged the hole with the boulder, and that was it. Absolute silence was everywhere, not a bird chirping, nothing.

To this day the spelunker thinks he had a concussion. His wife thinks it was all a dream, but there is the picture! Returning to the cave to try and recover their friend's body, they were unable to find the room they were in when the quake hit. On returning to the outside, they were also unsuccessful in trying to locate the opening plugged with the boulder.

Is there a city of giants underground? Was the big fellow from a subterranean world, irradiated by uranium ore? Would everything there be proportionately larger? Was that fellow guarding the gateway to the surface?

STINGING INSECTS AND GIANTISM

We now know that insects and rodents are very resistant to radiation. Perhaps they possess dormant genes from the beginning of creation. During that time everything was born on what to us would be a gigantic scale. Archeology and geology have shown us that such was the case. Radiation was heavy in those days and perhaps to the Creator, the beasties and bugs were tiny things compared to him. If those dormant genes were stimulated by radiation of the proper wave length or strength, the world might just return to giantism! A new heaven and a new earth might just have to start all over again!

Stinging Insects

The year 1979 was the worst in memory for bees, hornets, and other stinging insects. Wild wasps or hornets and yellow jackets were especially bad. They are meat eaters and are attracted from far off by food scraps and picnic lunches and open garbage cans.

Experts say the long hot, dry months were responsible, and the insects didn't go away until cold weather brought on the hibernation period. It seems to me that their numbers have been building up over the past several years, each year getting worse. (They seem to be getting bigger each year also.)

They have become not only a serious hazard to outdoor recreation, but a major menace to fire fighters on the various big burns during the summer. Stings are bad, but for people with allergies, multiple stings can be fatal.

The best medicine is prevention. Avoid nests that may be located in trees, under eaves in buildings, or in burrows underground. They may be in unlikely places, such as in a rolled-up sail on a boat. Swarming bees seldom sting but they can sure scare you. Fortunately, these insects hive up during the dark hours, so there is little danger from them at night.

Incidentally, if you spot a little round hole in your yard with the grass cleared away from it for an inch or so, this will be yellow jacket entrance to the hive, and there will be a guard or two at the door. Keep your distance! To destroy the hive, wait until dark, or better an hour after dark. Take a long-neck bottle of about a quart capacity and fill it with gasoline. Stuff the long neck into the hole but *do not* light it! By morning, all in the hive will be killed by the fumes. Wasps are a greater problem. Fire seems to be the most effective way to eliminate their nests. Fire, though, can be a hazard, so handle with care! There are sprays made just for wasps and bees. Other insect sprays will not work; they only make the insects mad as hornets! Don't spray around food or in enclosed living spaces; the spray can be as hard on you as on the insects.

For stings, carry in the car or have handy in the home a can of meat tenderizer or baking soda. Sprinkle it on a sting immediately. It breaks down and neutralizes the venom, which is a protein. Mrs. Robert Burlingame of Milton-Freewater, who has had previous experience with insect stings, says that rhubarb juice is also good. "We use the juice from the stock of the rhubarb. There is no waiting for the pain to leave. A dab of ammonia also helps," she said.

Like the armored knight with drawn sword, we may need such gear to survive, if insects grow suddenly to the comparable

FIGURE 46

We may need gear such as worn by the armored knight with drawn sword to fend off attacks of giant insects. *(Drawing by Pat McLelland)*

size shown in the illustration! Let's assume that ants would also become giants. In areas where the neutron bomb was used, they would have a field day. Thousands of people turned to mounds of jelly, scattered all around. What a smorgasbord for giant insects!

SUPER TERMITES, SUPER SNAILS, GIANT TOADS AND RATS

SUPER TERMITES INVADE SOUTHERN FLORIDA

Gainesville, Fla. (UPI) — The dreaded Formosan "super termite" can eat a house in six months, and scientists say it can't be stopped! [Bah!]

The termite, which gnaws through plaster, mortar, and wood preservatives to get to edible wood, has been discovered in south Florida, a University of Florida entomologist said Monday.

"There's no way to stop the spread," said Dr. Philip Koehler. "They've given up on trying to eradicate it. The only thing you can do is just wait for it to come." [Bull!]

The termite species, which belongs to the genus Coptotermes, was identified Monday by scientists with the Smithsonian Institution as the Formosan termite, one of the most destructive in the world. [Hog wash! They have to breathe don't they? God gave man dominion over all the earth!]

"It's been a great problem," said Rose Weck, a resident of a condominium complex at Hallandale near Miami, whose clubhouse roof was destroyed by the wood-eating insects. "These are termites with a vengeance." Unlike U.S. subterranean termites, the Formosan termite secretes an acid that allows it to burrow through plaster, mortar, creosote and nonedible materials to get at the wood it hungers for.

Koehler said it is a little bigger than U.S. subterranean termites, but about the same size as the drywood termite. Its wingspan is about three-quarters of an inch and its body about a quarter of an inch and growing!

The Formosan termite has been sighted in South Carolina, Louisiana and Texas, but this is their first appearance in Florida. The termite came to Koehler's attention when residents of the Hallandale condominium went to their county extension agent for help after three years of unsuccessful attempts by pest control firms to eradicate them.

The Formosan termite is resistant to most pesticides and even after double pesticide concentrations somehow it manages to survive, according to Koehler.

SUPER SNAILS ON RAMPAGE

Baracaldo, Spain (UPI) — Supersized snails are rampaging through plants and garbage at night and becoming a "real scourge," officials say.

Two of the super snails, each weighing half a pound, were discovered in a field in this northern Basque area by two youths during the weekend. The snails, approximately 3 years old, are the first to be found in Spain.

Giant toads are spreading northward from the Miami area, a University of Florida zoologist says. The toads, some of which reach eight inches in length, escaped from a shipping crate on a runway at Miami International Airport, about thirty years ago.

Evidently radiation is food for frogs also, as of August 16, 1980, down Miami way at least. Our giant blue herons wiped them out when they were imported from Africa and planted in the Willamette River in Oregon some years ago.

The American horned lizard or the Australian thorny devils that eat only ants might handle the Formosan termites! (See Plate 11.)

Giant rats on Bikini were discovered when an inspection crew landed to determine the effect of radiation on the plant and animal life there. There was need to know what forms of animals and vegetation were stimulated or killed by radiation. The rats seemed to be the only animal life that survived. They were classed as giants and found to be cannibalistic! The rats in the Philippines and Vietnam now average twenty-five pounds and are edible! The life forms stimulated by radiation may give humanity a life and death struggle for a long, long time! (See Fig. 47.) If there are any humans left on earth after an atomic war, those with the genes of giantism still in them might survive extreme radiation. The average young man now living on the so-called junk food is on average two inches taller than his parents!

SUPER RATS IN EGYPT
by Don A. Schanche

Los Angeles Times-Washington Post Service

Fisha, Egypt — An invasion of rats has ravaged Egyptian agriculture during 1979-80. Here rats weighing as much as four pounds each have stripped grain warehouses and entire fields of produce. Waves of tree-climbing rats stormed orange groves and devoured all the fruit. Roof rats, secure in the dried thatch that covers the mud brick houses, descend after dark and attack children!

RATS SCURRY FROM BRUSH FIRES

San Bernardino, Calif. (UPI) — All but one of Southern California's devastating brush fires was contained Monday, but authorities warned of two new dangers—disease-carrying wild rats and winter flooding.

The starving rats moved down the charred slopes around San Bernardino during the weekend to search the wreckage of fire-ravaged homes for food.

U.S. Forest Service officials said winter rain come could send floods rolling down the hills. That could prove as devastating as the fires, which killed four and caused over $70 million in damages.

Police shot dozens of the wild rats during the weekend. The hungry rodents had left 140 square miles of still-smoldering high desert brush and mountain timber in the San Bernardino National Forest.

"There's some real big ones up there," Sgt. Brad Hilder said. "We're concerned with two areas—disease and injury. . . .

GIANTS OF THE BIBLE

There were Nephilim (giants) on earth in the early days, and even afterward in Noah's time (Genesis 6:4). Noah lived to

FIGURE 47

Giant rats, weighing four pounds up to twenty-five pounds are found in many places. They are often used for food. *(Drawing by Pat McLelland)*

be 601 years by the time the flood dried up; he lived 350 years after the flood. God increased Moses' stature so that the people feared him (and I don't mean his prestige; he had that already). God increased the stature of two other men: "The Lord magnified Joshua as he did Moses (Joshua 4:14). "God enlarge Japeth (Genesis 9:27).

When Moses sent men out to reconnoiter the Promised Land, they returned after forty days and reported that "every man we saw there was of enormous size. Yes, and we saw giants there, (the sons of Amak, descendants of the giants). We felt like grasshoppers, and so we seemed to them" (Numbers 13:32-33).

Again in 2 Samuel 21:15-22, when David went with his guards to fight the Philistines, there rose Ishbibenob, one of the descendants of the giants. His spear weighed three hundred shekels of bronze (94 pounds). He was wearing a new sword and was confident he could kill David. Abishai, son of Zeruiah, went to David's rescue; he struck down the Philistine and killed him. Then Sibbicai of Hushah killed Saph, another descendant of the giants.

FIGURE 48

The Bible records many instances of men of enormous size. In this artist's portrayal of Moses addressing the Israelites, he is a man of huge stature: "The Lord magnified Joshua as he did Moses" (Joshua 4:14). *(Drawing by Pat McLelland)*

FIGURE 49

During World War I, men on both sides fell to their knees in prayer when the frowning face of Jesus appeared against a background of clouds. *(Drawing by Pat McLelland)*

Again war broke out at Gob, and Elhanan killed Goliath of Gath whose spear was like a weaver's beam. In Gath there was a man of huge stature with six fingers on each hand and six toes on each foot. He too was a descendant of the giants. These four, all giants, fell at the hands of David and his guards.

A GIGANTIC JESUS REVEALS HIMSELF AT THE END OF WORLD WAR I

World War I was mutually deadly, destructive to both sides and the world in general. In 1918, after a raging battle with thousands of artillery cannon chewing up the landscape and people, the entire front—soldiers on both sides—became suddenly quiet, not a sound. A white soft light spread over the terrain and in the sky around the clouds. And in the clouds there appeared the form of Jesus! This was not a cloud in the shape of Jesus! Men on both sides came out of dugouts and fell to their knees in prayer, all along the front lines. Cameras registered the picture. There for all to see was the face, head, and shoulders of the Lord Jesus, looking down very sorrowfully on the carnage! The form as it appeared in the clouds would be that of a person thousands of feet tall! Perhaps because of such savagery of man to man, he revealed himself to show his disapproval. Was this the Second Coming of Christ? "It is he who is coming on the clouds. Everyone will see him" (Revelation 1:7). With God, a thousand years is as a day.

World War I was not Armageddon as many thought it to be at the time. World War II came along later and proved to be far more destructive to life and property and the ecology. This war also was not the final one. Mankind is well on the way to Armageddon, though, and you had better believe it. There is no such thing as an atomic war with limitations. Whoever appears to be losing will go all out with no holds barred to win and, folks, that will be IT!

8

Armageddon?

THE GERMAN BEAST

Under Hitler, people were numbered, not on the head or hand, but in the armpit. This prevented harassment of the servants of the beast (Revelation 13:11-17). The first beast had caused a near-fatal wound, but Germany recovered from World War I. The second beast, Hitler, was allowed to work great miracles, even to calling down fire from the heavens on the earth. This was World War II. These miracles it was allowed to do on behalf of the first beast; it was allowed to breathe life into the statue so it could speak (radio propaganda).

FLASH POINT

Killing fallout is already in the atmosphere, placed there by over 900 atmospheric bomb tests! Debris has been falling out since the first test was made. This is the military and industrial complex of world powers playing with gigantic firecrackers, like kids. Each power has enough atomic weaponry to destroy the others many times over. There are three tons of atomic warheads for every man, woman, and child on earth! Without even having a war, there may be a great loss of life from fallout alone in the course of time, from what's already up there now! The metallic junk up there is going to come down eventually, and it may take a considerable toll. Radiation poisoning might be diagnosed as flu!

Russia is well aware that no one can win an atomic war. The United States is also cognizant of that fact. China, where life is as cheap as anywhere in the world, is now an atomic power. *They,* the Chinese, are still testing in the atmosphere even though the others have quit. *They,* the others, found out it was too dangerous to themselves, as well as the rest of the world.

The world's atmosphere *must* have a flash point. Thousands of atomic bombs detonated at once might just do it. Of course, such a kick might be enough to boost the earth out of its orbit! (See Plate 12.) Even if no flash point is reached, a mass detonation will cause enormous tidal waves wherever there is water on earth. Earthquakes will devastate the land. Russia has bomb shelters to accommodate thousands of people. Whole cities can be sheltered there. The real purpose, however, is to protect regiments of troops. People will be expendable! Animals, people, plant life, water, and air will be poisoned all over the world. The deepest jungles, the rivers and oceans will be contaminated!

What the results of an atomic war will be are there for all to see. There are none so blind as those *who will not see!* What a way to die, slowly cooking to death—an instant melt-down of all animal tissue from neutron bombs, or evaporation from the heat, like a drop of water on a hot stove! The Argus effect, as determined from the Ranier event of Operation Plumb-Bob, confines and reflects all radiation back to earth as a giant mirror. This shell is sixty miles thick and composed of beta particles completely enclosing the globe, and, according to the altitude of detonation perhaps 125 to 300 miles up, will be above the arc of trajectory of most IBMs. What's behind the button that can start the holocaust?

With the prevailing winds and jet streams moving around the world from west to east, if either China or Russia, or both, unloaded atomic bombs on each other and none of their allies lifted a finger to help them, even so, the blast effect would be awesome. The two belligerents would get the heaviest fallout from that day on for years. The deadly, silent, poisonous, burning of radioactive particulates, unseen—no sound on tin roofs—but interacting with the storm fronts around the world would be constantly pulled down by gravity. Every living thing would be deadly or dead—the air, grass, water, the very ground we walk on. No matter who strikes first, they both would receive the full benefit of the fallout, fourteen days more or less after the hit circles the globe. The maximum fallout occurs at the point of detonation the first time; after that, gravity and weather take over. Sweet revenge for those attacked even if they could not or did not retaliate. The fallout of radioactive ashes would fall on all—week after week, year after year, for who knows how long? Upper air storms would be brought down to earth with such destructive force that the earth might be stripped of all vegetation! *That's what's behind the button!*

Now let's assume—and that makes an *ASS* of *U* and *ME*—that some mad, power-hungry person or group in Nation A decides to strike first.

THE BEGINNING OF THE END?

One thousand pretargeted ICBMs are fired at Nation B. Nation B knows they are on the way. They fire as many or more, also pretargeted missiles. Of course each have anti-missile-missiles that can, as has been said, knock a fly out of the sky. Now if 50 percent of each salvo are destroyed in the atmosphere, that's going to be one thousand missiles of many megatons each that are literally going to blast our atmosphere into outer space! Air from the ground is going to be siphoned right out with it. The flame throwers used to consume the oxygen in caves, where Japanese troops held out in World War II, will be like a match in a hurricane compared to this! Hundreds were found dead without a mark on them, in cave after cave, suffocated. Although the military will all be more or less protected from blast effects, with the air suddenly pushing out into space, there will be no place that is safe. Missiles will be blasted off course far out and will, no doubt, collide with each other or other space junk that will be enough to set off further detonations. The telemetry will possibly be fouled up and many missiles may be turned back upon the senders!

Of course, the sudden release of air pressure on the ground will mean that animals, people, and anything with internal pressure of 14.7 pounds per square inch will explode. Ask any doughboy from World War I who saw a buddy disappear before his eyes when a high explosive shell landed close by. He wasn't blown to bits, he just exploded from internal pressure, into the created vacuum! You may have all the blast protection, all the radiation protection you can dream up, but the air in your body, the counter balancing internal pressure, will get you—just as in the Oregon Columbus Day storm when windows blew *out,* instead of in—only much, much worse. The thousands in bomb shelters and fallout shelters, those caught in buildings will be blasted out by the tidal wave of air leaving the earth for outer space. Of course, it will be a clean sweep because the bodies will vaporize, bones and all. No messy clean up! This without a bomb landing anywhere nearby!

Soldiers in tanks will simply disappear. All engines will slowly come to a halt for lack of air, or they may stop instantly if they are not air-tight. Only atomic-powered devices will survive; they will continue to operate in a dead world. *Great ships will wander aimlessly over the oceans.* Only atomic-powered submarines will survive but only until they surface, unless they are destroyed as they are cast ashore by oceans filling in caverns and depressions caused by earthquakes, air blowouts, or giant Sunami waves!

Earthquakes will split continents. The huge caverns created by underground bomb testing will explode, as will all areas where oil and gas have been taken out. What thin air is left will be saturated with sand and dust. The Sahara and other deserts of the world will be swept clean, down to the water table, or bed rock. The sun will disappear. The great sand dunes will be no more. Rivers will reverse their courses as the great Mississippi has all ready done once in the greatest earthquake ever to occur on the North American continent, in 1812. Steamboat captains have verified that!

Volcanoes will erupt all over the world. The molten magma will flow like water into the valleys. Mauna Loa, in Hawaii, the greatest volcanic mass on earth, had a lava flow that ran from the mountain for seventy-nine miles, to the sea, and continued to flow for over a year. It covered miles and miles of ground in width, pouring over old flows, adding many square miles to the Island of Hawaii, as it piled up in the sea. All of the islands were built in that way. With so much internal pressure, it might not take much for us to make a huge crack in the earth's outer shell! Mount St. Helens and others are only a pinhole of relief in comparison!

The aurora and other electric phenomena attracted to the magnetic poles will melt the ice caps. The oceans will rise hundreds of feet. Gravity will be so diminished that everything loose will fly high into the sky! The kick against the earth *will* nudge it out of orbit! GOD said, "I will shake the Heavens and the Earth shall move out of her place" (Isaiah 13:13). It will literally skip out from under all of civilization! Our beautiful water planet will have rings of debris around it like Saturn!

The moon, of course, will tag along. However, due to the reduced gravity, it will follow reluctantly, being left farther and

FIGURE 50

In a neutron nuclear war, all flesh will rapidly melt down. Tank crews will simply disappear. *(Used with permission of Wide World Photos)*

farther behind. When the united push from the blasts die down, unless we leave our solar system entirely, which isn't likely, the moon will see the earth go into a quick deep-freeze situation. Then it will catch up with a yo-yo effect. Being of the same polarity it will be repelled in time to prevent a collision with the earth. It will reach a gravi-pause situation (the distance required for perpetual orbit, where things are perfectly balanced, and will maintain it until the earth is slowly returned to its old orbit. "A new earth and a new heaven," all cleaned up (2 Peter 3:11-14).

If any animals survive, they will, by the grace of God, have been placed in the right place at the right time. When our atmosphere catches up with us and comes crashing down to earth, it will bring the extreme cold of outer space with it.

Attracted to the poles, it will begin reforming an instant deep freeze all over again, unless having passed through the fiery furnace that used to be the mesosphere warms it up enough so that in time the oceans will relocate on or in their former places, separated from the land, on the New Earth. With the oceans' abrasive action, the earthquakes will have leveled the highest mountains. The sands from the deserts will fill in the valleys; seeds will sprout, and life on earth will begin to multiply. Well, so much for prophecy. Let's see what's out of control here and now?

FIRE FROM THE SKIES

"The Day of the Lord will come like a thief, and then with a roar the sky will vanish; the elements will catch fire and fall apart; the earth and all that it contains will be burned up" (2 Peter 3:10-15). You ask, when is this coming? You hope it comes soon? There is one thing, my friends, that you must never forget: With the Lord, "a day" can mean a thousand years, and a thousand years are like a day to the Lord! The Lord is not slow to carry out his promises, he is merely patient with us, wanting nobody to be lost and everybody to be brought to change his ways. What we are waiting for is what he promised: The new heaven and new earth, the place where righteousness will be at home.

The fire coming from the sky *now* is not fire as we know it, but is silent, poisonous, hot, and deadly, and humanity put it there. However, anything within twelve miles of a bomb blast, depending on its size, *anything* that will burn will be ignited and burn! Unseen radiation will, like water, fill in all depressions. If the blast wave doesn't get you, something else will! These unseen things may cause the destruction of basic building materials, the mutation of animals, insects, and people. The poisoning of our earthbound oceans, the wilting of the human race, or, perhaps—giantism! What will those resistant ones, those who survive, look like? There are always survivors by the grace of God, somewhere, somehow. Insects do not seem to be affected by radiation! Volcanic *dust,* mistakenly called ash, as we know ash, is as fine as dry cement and just as heavy! It has killed innumerable birds, insects, and many other things that we are not aware of at this time, including fish. What's happening to bridges of concrete and steel all over the world? Is the concrete deteriorating, evaporating?

Are the malleable iron reinforcing rods affected by radiation?

Aren't bridges in other countries falling down, as well as many of our own? It's time we started testing our basic building materials for radiation exposure. *All research laboratories must be kept active!*

There is hope and possible salvation, or at least a possible postponement, for our earth as it is at present. The eternal optimism of youth, tempered by the experience of elders, can do it! The question-askers are constantly seeking the truth, in our great libraries, colleges, and universities. However, many times when they find the truth their superiors suppress it, because it's not good for the public, or it goes against the findings of another bureau. Biblical prophecies are happening!

We must make up our minds! Do we want a worldwide dust bowl?

Shall we go through the scorching, searing, birthing pangs of a new heaven and a new earth or shall we clean it up and keep it as it is?

EPILOG: CHRIST PREACHETH

In the meantime, when there were gathered together an innumerable multitude of people, insomuch that they trod one upon another, he began to say unto his disciples first, "Beware ye of the leaven of the Pharisees, which is hypocrisy.

"For there is nothing covered that shall not be revealed; neither hid that shall not be known.

"Therefore whatsoever ye have spoken in darkness shall be heard in the light, and that which ye have spoken in the ear in closets shall be proclaimed upon the housetops.

"And I say unto you my friends, Be not afraid of them that kill the body, and after that have no more that they can do.

"But I will forewarn you whom ye shall fear: Fear him which, after he hath killed, hath power to cast into hell; yea, I say unto you, fear him!"

And he said also to the people, "When ye see a cloud rise out of the west, straightway ye say, 'There cometh a shower'; and so it is.

"And when ye see the south wind blow, ye say, 'There will be heat'; and it cometh to pass.

"Ye hypocrites, ye can discern the face of the sky and of the earth, but how is it that ye do not discern this time?"

Luke 12:1-5, 54-55.

Appendix 1

WHY THE OREGON LEGISLATIVE COMMITTEE DID NOT RECOMMEND PASSAGE OF MANDATORY FLUORIDATION OF CITY WATER SYSTEMS

The hearing was held in Salem, Oregon, in the capitol building and was open to the public. The hearing of the pro's and con's was to determine whether or not to recommend the mandatory water fluoridation of *all* cities above a minimum population. The time of the hearing was to begin at 7:30 P.M. and was to end at 9 P.M. There were approximately 100 people in the hearing room, with a like amount in another room, standing room only, with closed circuit TV for them.

The people were there on time. They waited and waited. No committee showed up. At 7:45 the committee filed in, seated themselves, and the chairman rose to open the testimony. He stated that even though the hearing was starting late, the pro's would testify first and the con's could have whatever time was left; that when he pounded the gavel at 9:00 P.M., that would be the end of the hearing. This brought murmurs from the people and many snide remarks about a rigged hearing.

Several dentists testified, each giving his degree proving he was qualified to talk about teeth. Each practically parroted the other and each talked without interruption until they actually ran out of anything to say.

The chairman then called on a woman from California to testify for the con's, as she had come so far for the hearing. She stated that she was a former resident of Seattle; that she was constantly in poor health when she lived there. Her doctor, after making many tests, told her she had to move out of Seattle; that the fluoridated water was killing her. Her family moved to California, in an area where pure unmedicated water is used by the population, and her health has been steadily improving ever since. That ended her testimony.

The chairman next called on a professor from Oregon State University. This gentleman had walked in carrying a large shopping bag. He rose and went to the table microphone placed in front of the hearing's members. This was more like a courtroom than a public hearing chamber. The professor first stated his rank, name, and serial number so to speak. He asked the dentists if any of them had done any research on fluorides on their own. They all had not. They stated that all their information had been supplied by the American Dental Association.

As the dentists had all made much about their degrees, which they could rightfully be proud of, the professor stated that he also had similar degrees as theirs and many they probably never heard of! He also said he had done a great deal of research himself on fluorides and fluoridation; that fluorides had nothing to do with good teeth and that he had brought along evidence to prove it. He reached into the shopping bag and pulled out a huge skull! He turned it upside down and waved it around showing the committee and the people in the room. He said, "Look, here is the skull of an Indian I myself dug up in the southeastern Oregon area, hundreds of miles from any traces of fluoride. See what you are looking at, sixteen beautiful, perfect teeth!" He again waved the skull around for all to see, and they were beautiful, strong, solid teeth, perfect, without a single cavity or defect! He then placed the skull on the table and said, "Now see this," as he reached into the shopping bag and came up with the lower jawbone of that same skull. This he also waved around and said, "Look at 'em, sixteen more perfect teeth, not a cavity anywhere! Now this proves that a good diet is all that is necessary to make good teeth and bone." He left the skull and jawbone on the table to be examined by all who cared to. The gavel banged down and the hearing was over promptly at 9:00 P.M. Two people only got to speak in opposition, but what opponents they were to fluoridation!

Margaret Commack Smith, professor of nutrition at the University of Arizona, though not in attendance stated that "fluorine interferes with the normal calcification of the teeth, hurting the process of their formation, so that the affected teeth, in addition to being discolored and ugly in appearance, are structurally weak and deteriorate early in life." For this reason it is especially important that fluorine be avoided during the period of tooth formation, that is from birth to the age of twelve years!

Some doctors at welfare clinics are prescribing one drop of

fluorine in each bottle of a baby's milk; also, vitamins with fluorine in them, with no indication of the amount of fluorine prescription. They do, of course, say keep them out of reach of children. And that is also something else. The pharmaceutical companies and others have in some way managed to have the old red "Skull and Crossbones" removed from labels on items that are poisonous in any way, making it difficult for a child or those who cannot read English to know the contents are deadly. This symbol is as necessary as the international standards on the highways. It must be returned as a necessity right now!

I strongly recommend for all pro's and con's to read "The Grim Truth about Fluoridation, by Robert M. Buck; also, *The American Fluoridation Experiment,* by Exner, M.D., of Seattle, Washington, Geo. L. Waldbott, M.D., of Detroit, Michigan, and James Rorty, journalist (New York 1961). Both should be available at your library. Dr. Simon Beisler, Chief of Urology, Roosevelt Hospital, New York City, and past president of the American Urological Association, says, "It is now clear that fluoride is a potentially harmful substance when present in drinking water in *any* amount!"

One thing that is never mentioned is that the fluorine that occurs naturally in water and is only slightly soluble in water is calcium fluoride! The stuff put in drinking water is sodium fluoride. You heard about the rubber-legged cows at Troutdale, Oregon, that were grazing on pasture some distance around an aluminum plant! The owners won a court settlement and sold the land to the plant. And what about the forty-seven people killed accidentally at the Salem, Oregon, mental hospital. Someone, sent after powdered eggs for breakfast, evidently dipped into the roach powder barrel of fluoride instead! Aluminum companies are not allowed to dump their wastes into the Columbia River because the fish might be killed! But it is okay to refine the waste for your drinking water! Fluorides are cumulative, carcinogenic and deadly poisonous!

Appendix 2

WHY THE DEATH PENALTY

Let's tackle the terrorists first. I'm MAD, and I'm a polemicist of Holland Dutch ancestry, and a Methodist. Now that can be a *b-a-d* combination! In Holland the government was holding seven South Moluccans, belonging to a group seeking independence for their country. This group did not hesitate to open up on innocent civilians with automatic weapons, killing whoever happened to be handy. They then seized property and people to hold as hostages, to bargain for the release of their buddies from prison. After a sufficient interval to allow for careful planning, members of this same group seized a train in Holland. Again several people were killed. They demanded release of their buddies. Dutch commandos attacked the train and killed and captured some of the second group of terrorists. The captured terrorists are now also being held in prison. In a third attempt by members of the same group a consulate was seized, again with the spraying of automatic weapon fire all over, killing several civilians and, you guessed it, with the demand of the release of all the Moluccans. Again Dutch commandos killed several of the terrorists and took prisoner those who surrendered.

HOW CAN WE STOP THIS? If the first group of terrorists had been executed on the spot, the second attempt would not have been made. There would not have been many good Hollanders murdered! There would not have been an objective. These terrorists will not give up as long as there are live prisoners to be rescued. The Dutch government, by raising the ante, is encouraging the radicals to make another attempt. As more prisoners are taken, there is more incentive to free them. There will be #4, #5, #6, or more, until they finally come up with an irresistible scheme. Prisoners will then be freed with apologies, AND TERRORISM WILL PREVAIL ALL OVER THE WORLD!

Is it okay for radicals to kill good citizens but wrong for good people to eliminate those who will not live within laws necessary for the survival of society? People and governments are fed up with and strongly resentful of the namby-pamby do-gooder, the permissive, pseudo-Christian attitude of such people around the world! Must we warehouse thousands of murderers who never had it so good as they do in prisons? Who will eventually, in most cases, be released, if for no better reason than to make room for more murderers? Those who violate the death-penalty laws are well aware of the consequences. That is a universal language understood around the world! They are deliberately committing suicide. They are actually asking in their own cowardly way for society to do it for them. Let's do it at the scene of the crime!

Crime and Weapon Control

It's not the weapon you have, but the way you use it that defines a crime! If our government is sincerely trying to control crime instead of desperately trying to disarm citizens, if Congress has the guts to write a death-penalty law with teeth in it, we can control crime and weapons! And good Americans can keep their guns!

One of our Oregon senators said on gun registration, "I don't mind registering my car, why should I refuse to register my guns?" Of course cars are used to commit crimes, but they are seldom used as a weapon. When someone is trying to break down a door of your home, or crash in through a window, even though the intruder knows you are in there (and this has happened), would you rush into your garage to get your car to defend yourself and family with it? I guess that answers his question!

Of course, registration in itself sounds okay, but it's the first step toward confiscation by power-hungry, autocratic governments. Remember Austria, Czechoslovakia—they had gun registration. The Gestapo knew just where to go!

Remember how helpless the people of England were after Dunkirk? Until that time they had no need for weapons other than hunting guns, or those normally found in the kitchen! The bobbies didn't even carry guns! American sportsmen and women sent rifles and ammo by the hundreds of thousands to the English people, to enable them to put up some sort of resistance to an expected invasion by Hitler's blitzkrieg! We tolerate the Nazis, even protect them. It can happen here! It would, however, be much more difficult with an armed, well-trained citizenry. Gun control by itself or registration won't work. Any law officer knows it has been against the law for decades for a convicted felon to carry a weapon, and it hasn't slowed them down at all! Many people make a career of crime! Some have even organized it into a big, self-protecting business! They own judges, lawyers, and even have their own militia (hit men and women). By harassment, they can win a life sentence (seven to ten years, or less) for any criminal faced with a mandatory death sentence. That has been done in the South, in the so-called death belt! One prosecuting attorney even burnt his law books in public as a protest against such tactics.

How do we know a death-penlaty law will work? It's been used in Europe for years and years successfully! Offer a copy to Congress and you will find the only opposition is organized crime and its gullible dupes. The so-called humanitarians with Christian dissertations and legal gobbledegook, often disguised as non-profit organizations, out to make a buck any way they can, champion ghouls, killers, and murdering children (at 6 feet, 190 pounds, are fifteen-year-olds children?).

Recently, you may have received in a mass mailing a letter from the Julian Bond Death Penalty Project of the Southern Poverty Law Center, soliciting funds. Enclosed in it is a reprint of the *Chicago Tribune* of Friday May 27, 1977. Talk about organization! On the front of the envelope there are three pictures and, in large print, the heading: THEY DON'T EXECUTE CHILDREN, DO THEY? My wife asked me if I would want my boy to be executed if he had done what some of those did. One had stolen gas from a neighbor, threw it on him, and lit it! I said to her, "Our children know right from wrong. They must obey the laws, even as you and I." The age of responsibility should drop from eighteen to twelve, puberty! Execution every time and soon the so-called children would get the message—and so would the Fagans who use them to do their dirty work. They operate on the assumption that as a minor they can get away with it, and they do! Sometimes, of course, it's their own idea and that's a problem.

You, or some of you, may ask, why the death penalty?

Because THAT'S THE SOLUTION!

Long before Columbus discovered America, Europe had drastic methods of execution to deter crime. Where better than there to research law and the criminal? Torture, starvation, garrote, guillotine, hanging, keel-hauling—you name it, it's been tried. Many people do not fear death if it's quick and easy!

The stocks, the ducking stool, the calf put to the child molester and rapist, amputation of the hand that steals, and others, many others, were severe discipline or punishment. The punishment was designed to fit the crime, to deter. Most were cruel and very deterring—probably why there were more beggars than thieves? Human nature hasn't changed much over the centuries; perhaps a giant step backward would help us.

The then Governor Ronald Reagan in a newspaper article of April 23, 1977, stated that "from 1930 until the mid-fifties, when the death penalty was common, with several hundred executions each year, murder actually decreased, from 10,500 in 1935 to less than 7,500 in 1955. There is strong evidence that the death penalty as a deterrent to murder and crime control is successful."

Early in the 1960s an all-out campaign against capital punishment began, and by 1968 there was none. Murderers were not paying with their lives any longer. They were, however, making more people pay with theirs—12,500 in 1968, 18,500 in 1972, and more than 20,500 in 1975, an increase of 272 percent in twenty years. Society can't afford that kind of inflation. Those who say execution doesn't deter are very poorly informed, to put it mildly.

The chaplain in a large, old prison, when questioned regarding the execution of murderers, stated, "When it is unbelievable that they would regret their action or change their life style, a repeat performance is what must be prevented! On this premise only is the sentence of death legitimate."

We are warehousing thousands of murderers all over the nation! In these warehouses we also stock killers! There is a difference. Those who kill for fun, and professionals who kill for profit. A for instance—overheard in the locker room—"So you just got back from Nam?"

"Yeah."

"Me too."

"Oh, I've been back about six months."

"Did you kill many people?"

"Yeah, lots."

"Did you like it?"

"Hell, no. I was fighting for my life."

"Well, I liked it—sort of like a shooting gallery. Even now

after six months, sometimes I get this urge to get a gun and go kill someone, anyone. Since I joined the spa however, I'm rapidly losing the urge. These guys are a fine bunch of fellows, so different from the army. Even the muscle men are gentlemen."

One of these two is like any of us, capable of a murder in a rage, but less apt to, except in self-defense. The other is a potential killer—for fun, a sniper! A possible "for-hire" killer. Some charged with murder kill all witnesses while on parole or out on bail. These predators of the human race never had it so good. The law protects the criminal. *He's got rights* but, apparently, no obligations to society. The police are forced to tell him that he better shut up or he may incriminate himself! The organized crime lobby has now effectively prevented search of a vehicle without a warrant. If you want some one killed, shop your nearest prison for a repeater up for parole. The habitual criminal is sentenced for life (seven to ten years). The murderer is also sentenced for life, ha, ha, again (seven to ten years). Nothing funny about that, is there? But that's the law! Or is it the Parole Board?

It's no wonder that people have so little respect for law and are ready all over the world to reinstate the death penalty for those convicted of murder. When a federal judge is booed at a convention of judges, because when asked who is to blame for so much disrespect for law, he answered, "We are," it's time for lawyers and judges to look in the mirror and do something about it. Is it not true that the letter of the law only covers one type of incident? That the intent of the law is the real source of justice? The law is intended to be a deterrent and create a fear of discipline or punishment. It worked for Nebraskans.

Nebraska's right-to-kill law, so-called by its governor and passed over his veto by its unicameral legislature (one house should be enough for any state), worked extremely well during the time it was in effect. Seven burglars were caught in the act and killed the first night it went into effect. Crime dropped to almost zero. Criminals left the state in droves! Those well-established as residents trod the straight and narrow. Trigger-happy citizens were stacking them up. In the course of time, as rapidily as possible for a court to move, the Nebraska supreme court ruled it unconstitutional. It was unfair to criminals! The court stated that "it gave the citizens the right to be judge, jury, and executioner." Few lived to reach trial if they ran when ordered to halt or be fired upon. Fear of the consequences, not fear of the law, brought law and order back to Nebraska. Crime got the message. It went on strike! Who fears the law? The police and you and I, that's who. To date it has been impossible to get any data on how effective the law really was in Nebraska! I wonder if the right-to-know law would work?

It costs $10,000 per year to keep the worst or meekest con in jail for a year, according to Governor Straub of Oregon. It costs only $6,000 per year for a good honest Jack or Jill to go to college, with plenty of fringe expenses in addition. Any work available on the side is also a big help. Work, however, is a dirty word to the criminal and many of the younger generation, to say nothing of hard labor—man, are you crazy? That would be an excellent deterrent for all lesser crimes. A year or more on the road gang or making gravel at some quarry supervised by the state would really make lesser crimes unpopular. Whatever became of the hard labor sentence for crime? Did organized crime get rid of those judges also? Of course Jack or Jill can qualify for scholarships. There are also student loans available, which must be paid back, of course. There is, however, another way known to the scofflaw (that word originated in Prohibition days to describe gangsters, racketeers, and bootleggers, many of whom were armed with tommy guns. What could police do, armed with a .38 caliber revolver? Someone won a $1,000 reward for coining it. Too late, however, because of repeal, to serve its purpose, or was it?)

Now about this other way. What's the incentive to go that route? No registration fee, unless committing a crime can be classed as such. No tuition and free room, board, medical care, small change for incidentals, and the only work will be to learn a trade if you don't already have one. Or you can study for a college degree in numerous professions, with no guarantee it will keep body and soul together. Sometimes the supervision and associates are a bit rough, but what do you want for nothing? (It's just like a hitch or two in the Armed Forces, without the freedom they have, depending on how much you value your freedom. Of course, you may get an honorable discharge from them, which gives you free burial in any G.I. cemetery, a nice fringe benefit nowadays. Also, if you ever reach the down-and-out state, you can always turn in to any army installation for a free meal or so! Keep your discharge handy.)

In this for-free route, most of your thinking is done for you, with fringe benefits such as connubial visits from or to the wife or girl friend. They no longer turn cons loose, as they used to in Oregon, to visit with a female! Carl Cletus Bowles, the convicted murderer, killed that practice, along with several innocent hostages. Execution would have prevented those killings, saving several good taxpayers, as well as a father and mother. The cons in Oregon State Prison would like to get him back; he's dead if they do. As it is, he will escape or be released again to repeat over and over, supported by taxpayers' money, yours and mine, to the tune of $10,000 a year. That school isn't for just anyone.

To break into that college, it is necessary to kill someone, or get seven to ten somehow. This is your entry fee. Of course, it will mean seven to ten restricted to the campus with time off for good behavior, at least for a while. Of course, there is work-release, but that will cut into study time. You surely can earn your degree in that length of time. If you act like a fool and are a model prisoner, when and if the parole board calls the court to find out how many to turn out to make room for how many more according to their evaluation charts, it really doesn't make any difference what sentence the judge lays on, you may get the boot. Then you will have to violate parole or kill again to be readmitted to finish your education! Sometimes the warden is very inconsiderate!

Before Oregon became a state, it had in its territorial constitution the death penalty, for stealing a horse and for murder. Not just for shooting a policeman (our finest), or killing a man, a woman, or a child, except in self-defense. If you used poison, a gun, a knife, or cast-iron skillet, no matter what, you'd had it (this was very nearly equal to the old English law). This controlled murder very well. The death penalty remained until capital punishment was abolished in 1914. There was not much crime up to that time as far as murder was concerned, so who needed it? Things went from indifferent, to bad, and then to worse and worse, and after six years in 1920, the people reinstated capital (mistakenly called) punishment. When a suspect was proven guilty of a capital crime, the sentence was death for the guilty in any manner prescribed by law. Definitely not punishment but elimination. How about that? The criminal element got the message at once, and for forty-four years murder wasn't worth the price.

With things getting better and better, with less and less crime, and chances of apprehension greater than ever, the legislature in

1964 offered a constitutional amendment to the voters. This in answer to a petition to Governor Hatfield in December 1963 from the Greater Portland Council of Churches to abolish the death penalty. Hatfield, apparently standing as far as possible from the position of spineless pity in which our humanitarians take such pride, replied in January that "it was a matter for legislative action and that it was up to the legislature to consider it." They did and it passed. The deterrent to murder was removed and replaced by an incentive! The killer and murderer were given to understand, indirectly of course, that they were the jewels that must be rehabilitated and preserved for posterity. They were to be, unknown to good Christians, the future leaders of a generation of children educated in every crime in the book. The children had to live somehow. The three R's were given the heave ho by educators. School dropouts became runaways. The ultimate tool of organized crime was created.

Now, teenage male prostitutes (chickens) to use the argot, openly solicit trade in the big cities. Crime is rampant again and organized like business, big business! A person is not safe on the streets anymore. Gangsters, (scofflaws) often operate their business from inside a prison. Censorship is forbidden. A released or paroled inmate, or a work-release trusty, can be sent out as a hit man or enforcer! Censorship would prevent that, but of course it would interfere with their rights! Their lawyers have also acted as couriers.

Of course, no murderer has ever been deterred by the death penalty! If he were, it wouldn't have happened, would it? Who can say a law isn't a deterrent until he has talked to those who have been deterred by or because of it? I'm sure there are lawyers, or those in the penal system and law enforcement division, or psychiatrists—the last being credited with having an open-sesame to the human mind—who have considered killing someone but didn't because of a law against it! Like the housewife who, when asked if she ever considered divorcing her husband, replied, "Divorce him? No. Murder, yes." Most of them were deterred.

You can find someone if you try! The threat of death for burglary, theft, or you name it in Nebraska was a real deterrent, wasn't it? It also worked in England! I know of people who have said, "I'd kill the S.O.B. if it weren't for the death penalty" (when it was in force here), or, "I'll wait and see how the trial comes out." The do-gooders ask, who will throw the switch or drop the pellets or spring the trap, will you? Just invite the mother, father, brother, or sister of a girl who has been gang-raped and beaten to death afterward. Who needs paid executioners?

Why do killers kill? Our military has trained soldiers in how to kill a man quickly in nine different ways, barehanded. It's very difficult to make an arrest when those fellows resist. Two deputies tried and failed, until aided by a passerby who also was trained in the nine-way method. The only force they respect is threat by a gun. *That* they understand. Then, because that method was too slow and dangerous, the military has trained millions to do it wholesale, with the finest weapons man can devise. These people are carefully screened before discharge, but some slip through and, unknown even to themselves (and *you* may be one of them), carry this deeply imbedded urge to kill. When and if it hits you, you'd better suppress it, buddy, for the death penalty is coming back, and, remember, your hands may be considered a deadly weapon! It would be a stinking shame to live through Nam, and then commit suicide here.

Some people are triggered in different ways. A wife slipped up behind her husband and goosed him. He, recently in combat and discharged with a disability, whirled and hit her on the chin, knocking her clear across the room and out cold. She never did that again. He explained to her that combat began with a touch! This trained killer has simmered down and raised a fine family.

Unknowingly, I witnessed the triggering of a premeditated murder. Several years ago, while parked in a car on skid row (the wino district of Portland), I watched a man walk up in front of a pornographic theater and stand straddle-legged in front of several life-sized pictures of young women, displayed on the sidewalk, in flimsy negligees, spread-eagled in positions of torture. This fellow stood staring at the pictures for over five minutes. The ticketseller (a man) was also watching. The fellow standing so also attracted other watchers. He hurriedly left when he became aware of others watching him.

Premeditated murder occurred a few weeks later. Two Oregon girls, hitchhikers, were the victims of this over-sexed maniac, one at a time. Perhaps the first one died too fast! The women were found hanging in the garage of a man in Salem, Oregon, spread-eagled, in the positions of obvious torture, as shown in the posters, and his wife claimed to know nothing about it! It is possible, however, that she was the tip-off. Would you warehouse him for seven to ten years and turn him loose? That is what will be done, no doubt, with the almost certainty that he will repeat! What the hell's the matter with us? Parole boards, psychiatrists, do-gooders, and sobsisters have killed more innocent people by trying to change the stripes on the tigers. They should be held responsible for repeaters, if this practice continues. Or the people may go the route the Nebraskans took, without the benefit of legislation!

You have heard that the English bobbies don't carry guns. People who use that line as a clincher don't tell you that *that* was a long time ago. They honestly don't know why they didn't carry guns. No one I ever asked knew why. *Now,* since the repeal of the death penalty, according to a British consul, they do have guns within easy reach, perhaps like the shotguns our police carry for self-defense? Their crime rate will soon exceed ours. Recently they missed reinstating the death penalty by a very narrow margin. They (the officials) want the old law back. Why? Because it was a deterrent of crime.

I obtained the following information from the British information service. "Under their old law, if a *weapon* was used in the commission of a crime, the criminal was tried under the death penalty statute." This was not meant to be punishment but *elimination* of the criminal from society, and it almost did just that. Vicious crimes were few and far between. To quote a British consul again, "If a *weapon* was used, the criminal was *very* careful to use one that was easily disposed of." No wonder bobbies didn't need to carry a gun. Organized crime has since become so strong in England that with their propaganda and clever use of dupes and gullible people, they were able to prevent the return of the old law. But by the next time around, things should be bad enough to convince the good citizens to bring it back.

Don't we have the guts to write a law like that? A two for one law—weapon and crime control—all in one package? The use of a weapon in the commission of a crime to be punishable by death! Who will need new prisons? Such fine hotels are actually desirable to certain elements of our society. Some have been known to deliberately commit a crime so they will be sent back! Some well known to authorities are just jailed for the winter, which is what they want anyhow. They never had it so good in civil life, which doesn't say much for our society. They simply can't cope with life on the outside. A sentence to the military for lesser crimes would deliver to the Armed Forces the type of people that can fight, those who like to have their thinking done for them and need discipline. The lesser criminal needs security and often becomes a career person in the military. Some judges go that route.

A recruiting sergeant told me he stole a car for a joy ride. The judge said three years in the pen or a three-year hitch in the Armed Forces. He likes it. It's his career.

Let's recycle the vicious criminal and arsonist. Does not the Bible say, "Ye must be born again to enter the Kingdom of Heaven." Let's help them out. Better luck next time. Man cannot and should not be able to live outside a society, whose laws are necessary for his physical survival.

Prisons with walls thirty feet down in the ground and sixty feet above, with guards and guns all around, may soon be needed to keep people out, if the present trend continues. A crime college for the vicious is definitely not needed. New prisons are not the answer.

Science, which has taught us so much about killing, could at least teach us to kill decently—an anesthetic that would permit the condemned to pass from a state of sleep to death, within easy reach for some time for self-administration, or a cyanide tablet instead of a gas chamber (instantaneous), or carbon-monoxide gas, odorless, tasteless (from sleep to death). In case of loss of nerve or religious scruples against suicide, it could be administered in a cell. This would provide a little decency far removed from the cruel and inhuman punishment. Sort out the killers. Dispose of bad stock. No more warehousing. The *elimination* of the criminal, if that is what is required to make the world safe for people, must be done.

Remember, the *death penalty and weapon control in one package.* A simple and sure solution!

Appendix 3

EIGHTEEN IS TOO OLD!

Eighteen is too old for WHAT? It's too old to start learning to live, that's WHAT! Lowering the voting age to eighteen shows that some people are stumbling in the right direction—down, down, down. The sales pitch to lower the voting age was favored by nearly all politicians, it seems. Was it an effort on their part to lower it to a more gullible age? Most adults are well aware that skillful politicians can mold the thinking of our youth, like a sculptor does clay. People usually become less gullible with age.

Then there is that old quote, "If they are old enough to fight, they are old enough to vote." I've seen some ripsnortin' fights among kids under twelve. Try two in a playpen! Perhaps we should lower the voting age to twelve? Well, hardly, as I am sure this age would vote for ice cream to be served three times a day in school, but would *not* vote for twelve months of school, even with ice cream every hour on the hour as an inducement! Twelve years, however, is a good age to start teaching youngsters to qualify for living and voting, but first teach them to live, and they will qualify as voters at eighteen! Look up that word *qualify,* it's a corker!

Now *you* really want to help the young people? *You* want to stop the crime wave? I don't mean just juvenile crime, either. Let's institute a compulsory course *now* on how to live. Let's start at the age of twelve, in all schools, public and private, nationwide! Each school shall have an active truant officer, and not just the principal. School shall be compulsory, through twelve grades.

If you can't be bothered, and always take the easy way out, *this is it!* I feel sure people who would stop the world and show teenagers where to get off will like it too. It might help get the two types together, and wouldn't that be great?

Well here it is. Let's call it a Survival Course for twelve to eighteen-year olds—let's drop the word "juveniles." We must be honest and practical; it's as easy as being an officer and a gentleman. That's tough enough. Let's remove the coddling, protective laws from all youth age twelve to twenty-one. Just declare them adults for all practical purposes. Let all adult behavior laws cover these youth. Laws are only guidelines set up with a certain intent in mind to be interpreted by judges who, we *hope* and *pray,* are qualified to enforce the intent as well as the letter of the law. Let it be nationwide—*and now;* it's urgent!

Mother Nature's laws have been ignored too long. Children, at approximately age twelve, sometimes sooner, cease to be children and become youthful young men and women, whether the laws of man or mother or dad like it or not. This is Mother Nature's age of puberty. From this age on, all teenagers want the same thing! They want to grow up! What's their universal beef? "My folks won't let me grow up." Well, folks, there it is—the age to nail it down and start teaching them to qualify for living; the age that they want to start. At least no matter where or when they decide to strike out on their own, they will know that living properly is a serious and responsible business. That word "drop-out" is a denigrating word and is better forgotten. Dropouts all too often become runaways, turning into criminals, unwed mothers, or victims of the Fagans and Madams of the world.

Our youth have intellect running out of their ears. These minors are very intelligent and often better informed at twelve than high school graduates of fifty years ago. Their constant *why*? Their eagerness to learn, their desire to help when they are too small, their indifference when older and able is our fault. Can't we cope with this? How about mother staying home until the youngest child reaches twelve years of age, *if* there is a working father in the family? Aren't we smart enough to figure this out? These kids are soaking up knowledge like a sponge. They are in there digging, digging, all the time mining knowledge where-ever they can find it, good or bad, and bad is so much easier to find and *much* more exciting. If we use the word *miner,* instead of minor, it describes them perfectly. Schools educate them in everything but how to live and be responsible! Mother and dad are supposed to take care of that but, usually, after taxes, they are too busy trying to find enough to live, in the manner to which they would like to become accustomed, and they want it *right* *now!* But the curriculum *can be made* interesting and exciting. Park Rose High School, District 3, has done just that!*

There is *nothing* free anymore, not even the air! Our youth can take much in stride while they are youths, but must lay out this course in how to live, and we better hurry up with it. Can they learn in six short years, twelve to eighteen, how to *live* and survive, no matter when *they* decide the system has nothing more to offer and they strike out on their own, as they are doing now?

Some people think voting at eighteen is pretty risky stuff. After all, until youth reaches twenty-one, there will be many things they *can't* do. In most states they are not allowed to frequent taverns or buy whiskey until they reach twenty-one. *Why?* Because well, that's pretty heady stuff. Most folks think the teenager needs more aging, maturing and mellowing; like whiskey, it's not so good green, and there is *no* good whiskey, no matter how old it is! Fermented apples in the snow have made many a robin drunk and an easy catch for cats! Many people, and with good reason, fear the consequences of green decisions and impractical, uninformed votes, though there are plenty of those from people between twenty-one and senility. Youngsters want to know so much so soon, and there is so much of living that is not in books. Bless their little undisciplined hearts.

Discipline is another thing that, believe it or not, they are all begging for! Some even feel unloved because the other kids get licked, and their own mothers or dads don't give *it* to them, when they know they have *it* coming. They are smart enough to detect what is, from their viewpoint, indifference on the part of mother or dad to their problems. They strike out in resentment, and rampage around begging for discipline from adults that *know* they don't dare lay a finger on them! The kids are smart, and know that also. There is nothing that bugs an adult more than a kid, or a pack of kids, tossing out sarcastic or snide remarks under the protection of the law. The adult holds himself or herself in check, for fear of the consequences, and kids spot fear as quick as a dog, if not quicker. That protection has got to go! Remove it at the age of twelve! You will find then that "Yes, sir," "No, Ma'am," "Please," and "Thank you" will become

For a copy of *Course Guide, a Process Model for Studying Economic Elements of Contemporary Family Life,* by Cliff Allen, write to Park Rose High School, 11717 N.E. Shaver, Portland, Oregon, 97220. Enclose $2 for postage and handling.

popular fast. Not all problems will be solved without remonstrance, but the news gets around, and *hearing aids* will not be needed so often! Then there is always mother and dad to be reckoned with, if things get out of hand. The kids will have rights, under the law, even as you and I! And that is pretty lenient!

If those between twelve and twenty-one are not adequately protected as adults, then the laws are not adequate for adults! Let's start to cure this social cancer now. Instead of spending more money warehousing and carefully placing in isolation our most aggressive, nonconforming, undisciplined youth, let's try this. With firm handling and extra classes instead of suspension or expulsion, by keeping them *in* school, they could become some of our best citizens.

First should come compulsory physical education for the able-bodied, not a choice of a specific sport, but a plan of complete body conditioning for young men and women. A plan of exercise that could be used all through life would save millions of lives and keep the body and mind physically fit, mentally alert, and *morally straight!* This would tie in with the President's program timed to do the most people the most good. Athletes would be the natural offshoot of such a program. The first step in self-discipline would have been taken because they liked it! Those interested in special fields of athletic endeavor would gravitate in that direction of their own accord.

Second—all schools, public and private must have a disciplinarian! Not a teacher or principal but probably the physical education instructor, an athlete who is capable, ready, and willing, to discipline students as well as being a student of youthful behavior, but *not a psychiatrist!* One for the boys and one for the girls. Most youths respect physical fitness. Others *can* be taught to! A slap on the face by a teacher should suffice for class discipline, with the privilege to go as far as necessary to maintain order! Where real corporal punishment is indicated, let the physical education instructor be the judge to dispense as he or she sees fit. Perhaps the class boxing champion could go two or three rounds with a bully! Youth will get the message and shape up fast. In time, they will be grateful for it.

Some of us are worried about becoming a nation of hoodlums. We should be. We are rushing toward it as fast as possible. We take our finest potential citizens out and shoot them! Not literally, of course, but we offer them to our enemies, and if they are good fighters, they will take care of it for us. It's easier that way!

That perhaps is a good idea for the hell-raisers, but we send our best to war! Many of our enemies may be illiterate and poor thinkers, but they can be trained to be terrific fighters and they've *never* gone to school.

Our youth are smarter than the adults! They knew a police record would keep them out of the draft, so—up go car thefts all over the country. They outfoxed the law. Pretty soon, no army. You want to stop crime? Organize discipline battalions and watch crime take a nose dive. This group wants to *fight.* They are naturally aggressive. Who needs high paid professional armies? These folks will do it for fun! This type of youth may even make the military their profession?

Let's require the judges of our fifty states to sentence to hard labor anyone whose crime calls for six months of confinement. Those sentenced to a year or more would be required to enlist in either an army or marine disciplinary battalion. They would have their choice of either of these fine outfits, both of which are ready, willing and very able. They could shape them up and ship them out as good citizens! The only parole would be to be an Honorable Discharge, and all would be right with the world again! *Who needs a draft in peacetime?* The isolated, the foxy ones will become good citizens and find something to do in a hurry. *Hard labor, wow! Anything but that!* A useful citizen *is* made. Close the penitentiaries; the military can shape them up!

We must devise a system *now* for training the youth of our country. Those, including the adults, who cannot or will not live within the laws of society that are necessary for its survival must be isolated or *eliminated!* A rough, tough, system that will be *long* remembered. Twelve to eighteen years is a *very* impressionable age. Many achieve adult stature with the undeveloped mind of a child! This *is* a handicap and must be allowed for. Take a class to a large city or county jail or a state penitentiary. Show them the bull pen and the rock pile, and you will really be getting somewhere. They will see the handwriting on the wall and remember that visit forever. A trip like that is worth millions to law enforcement and good citizenship. Rough on the little darlings? Remember, you are trying to protect the adults from them, and at the same time trying to make them good citizens able to survive on their own.

Third and most important is *incentive pay* for good teachers and *disqualification* for those who just don't have what it takes! It's another must. Many students in both high school and college, according to an authoritative survey, gave up in disgust, due to the poor quality of instructors! A student from one of the top colleges in the state said to me once, "Mr. Clare, *please* don't blame the kids for everything, when it's really those *kooky* professors!" There is *nothing* more important than *good teachers!* They are, our *greatest treasures!* Good services don't come cheap.

Every state in the Union would benefit by making it mandatory to have a course such as the Economic Elements of Contemporary Family Life as part of the high school curriculum.

Legal marriageable ages seem to be a good place to start to find out who needs it the worst.

Boys may marry with consent in four states at fourteen years, one state at fifteen years, eleven states at sixteen years, four states at seventeen years, twenty-nine states at eighteen years, and one state has no minimum.

Without consent, boys may marry in seven states at age eighteen years, five states at age nineteen years, three states at twenty years, and thirty-seven states at twenty-one years. No consent for boys under eighteen years may be waived in two states.

Fourteen states have no waiting period; two states have waiting periods of one week; two states, two weeks; twenty states, three weeks; one state, four weeks; nine states, five weeks; one state, seven weeks; and one state, ten weeks!

In forty-seven states medical exams are required, but in eight of them it may be waived; five states do not require examinations. Common law is recognized in fourteen states.

Girls may marry with consent at age twelve in four states, in one state at age thirteen, in one state at age fourteen, in seven states at age fifteen, in six states at age sixteen, and in thirty-two states at age seventeen. At age eighteen, the girls are on their own in all states!